Springer Theses

Recognizing Outstanding Ph.D. Research

For further volumes:
http://www.springer.com/series/8790

Aims and Scope

The series "Springer Theses" brings together a selection of the very best Ph.D. theses from around the world and across the physical sciences. Nominated and endorsed by two recognized specialists, each published volume has been selected for its scientific excellence and the high impact of its contents for the pertinent field of research. For greater accessibility to non-specialists, the published versions include an extended introduction, as well as a foreword by the student's supervisor explaining the special relevance of the work for the field. As a whole, the series will provide a valuable resource both for newcomers to the research fields described, and for other scientists seeking detailed background information on special questions. Finally, it provides an accredited documentation of the valuable contributions made by today's younger generation of scientists.

Theses are accepted into the series by invited nomination only and must fulfill all of the following criteria

- They must be written in good English.
- The topic should fall within the confines of Chemistry, Physics, Earth Sciences, Engineering and related interdisciplinary fields such as Materials, Nanoscience, Chemical Engineering, Complex Systems and Biophysics.
- The work reported in the thesis must represent a significant scientific advance.
- If the thesis includes previously published material, permission to reproduce this must be gained from the respective copyright holder.
- They must have been examined and passed during the 12 months prior to nomination.
- Each thesis should include a foreword by the supervisor outlining the significance of its content.
- The theses should have a clearly defined structure including an introduction accessible to scientists not expert in that particular field.

Philipp Antrett

Characterization of an Upper Permian Tight Gas Reservoir

A Multidisciplinary, Multiscale Analysis from the Rotliegend, Northern Germany

Doctoral Thesis accepted by
the Rheinisch-Westfälische Technische Hochschule
Aachen, Germany

Springer

Author
Dr. Philipp Antrett
Energy and Mineral Resources Group
RWTH Aachen University
Aachen
Germany

Supervisors
Prof. Peter Kukla
Energy and Mineral Resources Group
Geological Institute
RWTH Aachen University
Aachen
Germany

Prof. Dr. Harald Stollhofen
Geo-Center of Northern Bavaria
Friedrich-Alexander University
Erlangen-Nürnberg
Nuremberg
Germany

ISSN 2190-5053 ISSN 2190-5061 (electronic)
ISBN 978-3-642-36293-4 ISBN 978-3-642-36294-1 (eBook)
DOI 10.1007/978-3-642-36294-1
Springer Heidelberg New York Dordrecht London

Library of Congress Control Number: 2013931943

Printed on acid-free paper

Springer is part of Springer Science+Business Media (www.springer.com)

Parts of this thesis have been published in the following journal articles:

Antrett, P., Vackiner, A. A., Kukla, P., Klitzsch, N. and Stollhofen, H. 2012. Impact of Arid Surface Mega-Cracks on Hydrocarbon Reservoir Properties. AAPG Bulletin, V. 96, No. 7 (July 2012), pp. 1279–1299.

Antrett, P., Vackiner, A. A., Kukla, P., Back, S. and Stollhofen, H. 2012. Controls of reservoir compartmentalization of an Upper Permian tight gas field in Germany and links to a modern analogue in the Western U.S. Petroleum Geoscience, Vol. 18, 2012, pp. 289–304.

What magically and rejuvenating qualities those dunes and canyons have; they heal the spirit and calm the troubled mind

—Writing on the wall at the Panamint Springs Resort, CA, USA

Supervisor's Foreword

Natural hydrocarbon gas is considered to represent the "bridging fuel" until new energies become technically and economically viable. Amongst hydrocarbon gas, one can classify conventional, unconventional, and tight gas resources. The latter two have recently received much interest because of the potential very large reserves, which could be produced with suitable technology. Tight gas reservoirs which are found throughout the world and which occur in all common types of reservoir have been produced for many decades, but still pose a major technical challenge owing to their heterogeneous reservoir characteristics and in particular their low permeability and low porosity. Given the global importance of such reservoirs, the understanding of the complexity of tight gas fields therefore requires an integrated approach involving geological, geophysical, and petrophysical analysis which very few publications to date have achieved.

The thesis of Philipp tackles a complex problem associated with the geological history of tight gas sandstone reservoirs in the Permian of northern central Europe. It presents an approach of integrated modeling, laboratory, and field work including reservoir properties, petrography, lithofacies and sedimentology, seismic attribute analysis, core analysis, and nano-porosity studies. Despite a long exploration and production history in this basin, previous work in Germany has mainly concentrated on large-scale basin analysis based on seismic and wireline borehole data and detailed diagenetic work based on core data. The Permian sequences in Germany occur at depths of up to 5,000 m and are characterized by low permeabilities owing to deformation, fluid flow and hence diagenetic alteration of the largely fluvio-eolian clastics.

With the background of the considerable global importance of tight gas reservoirs, this work presents a multidisciplinary approach to explain the still enigmatic evolution and distribution of these reservoirs and their properties. Aims of the study thus included a model of processes responsible for the loss of good reservoir properties during burial and subsequent events in the basin. Methods were used to assess internal reservoir structure and diagenetic patterns, to understand hydrodynamic controls within the basin and the sedimentary facies' which dominate the architecture of such reservoirs and to describe the pore space characteristics.

Initially, seismic and subsurface data were interpreted and analyzed and finally compared with a modern analog in Panamint Valley, western USA, where structural and sedimentary features including arid-surface mega-cracks have been mapped and confirmed by geoelectric measurements. Based on the integrated analysis, Philipp can conclusively prove that the interaction of syn- and post-sedimentary tectonics and burial diagenesis plays a crucial role in the evolution of deeply buried tight gas reservoirs. The reason for the low permeability of these reservoirs is manifold and controlled by a large number of parameters and processes, mostly of sedimentary and structural origin. Depositional controls, burial history, and associated diagenetic changes play the main role in the protracted development of the reservoirs. The role of authigenic clay minerals for example is one of the main reasons for the low permeabilities encountered. It is particularly the high-resolution characterization of the pore space, the importance of early diagenetic illite and chlorite coatings in sand dunes and the role of arid surface mega-cracks leading to subsequent reservoir compartmentalization which are a major step forward toward the understanding of deeply buried clastic reservoirs.

Philipp's research formed part of the tight gas initiative (TGI) between RWTH Aachen University and Wintershall Holding GmbH which supported this study. He has presented his work in several international publications and at international conferences. His thesis is a well-written and well-illustrated piece of work which provides a comprehensive model of the multi-scale Permian tight gas reservoir evolution in space and time.

Aachen, September 2012 Prof. Dr. Peter Kukla

Acknowledgments

I am deeply grateful to my supervisor Prof. Peter Kukla, Ph.D. for the opportunity to be part of the tight gas initiative. Thank you very much for your trust and your support during this project. My sincerest thanks go to my advisor Prof. Dr. Harald Stollhofen. It was always exciting discussing about sedimentology with you, even in the core lab (with our minds walking in the dunes of the Namib Desert). Many thanks to PD Dr. Christoph Hilgers for "taking care of me" during the last year. Your enthusiasm, dedication, and excitement really pushed me forward and motivated me during the last phase of my Ph.D. I enjoyed our discussions and brainstorming sessions a lot. My thanks also go to Dr. Frank Strozyk for his advice and support. I appreciated your patience, your positive criticism and persistence. I would also like to thank Dr. Norbert Klitzsch for introducing me to the world of geophysical measurements (at least to a tiny little part of it). Thank you for borrowing us all the expensive stuff in the big boxes and for being patient and understanding when DHL nearly made my head explode. Furthermore, I want to express my gratitude to my co-authors for the help with improving my manuscripts, my colleagues at the Geological institute especially to Beke Rosleff-Sörensen, Uwe Wollenberg for spending uncountable days at the SEM and to Conny Lutter for her motivation to dig up the literature, even during her holidays.

From the Wintershall side of my project, I want to thank Dr. Harald Karg, Michael Blum, Dr. Wolfram Unverhaun, Dr. Claudia Bärle, Bernhard Siethoff, Petra Unverhaun, Dr. Wolf-Dieter Karnin, Jan Himmerkus, Dirk Adelmann, and Dr. Dieter Kaufmann for many fruitful discussions. Thank you very much for your support and for being so straightforward. My gratitude also goes to Dr. Anton Irmen and Dr. Robert Bussert of the GDF Suez.

I am very grateful to Richard Friese and his colleagues from Death Valley National Park for granting our Research Permit and for their logistic support during our field study. Additionally, I would like to thank the staff of the Panamint Springs Resort, especially Rosemary who "adopted" us as her grand children, Ashley for making the best coffee in Panamint Springs, Alexis, Victoria, Aaron, and Uncle Bob. And of course, thank you all for granting us the family discount!

I thank my family, Alex family, and my friends for support, encouragement, uncountable care packages, and sometimes just for distraction. Last but not least, I want to thank Alex for uncountable geological fights, two unforgettable field studies at the Death Valley National Park and her patience when I reached one of the many motivational and spiritual synclines of this 3-year Ph.D. roller coaster ride.

Contents

Chapter 1
Introduction

This chapter describes a multidisciplinary, multiscale approach to the analysis of tight gas reservoirs. It focused initially on the facies architecture of a Permian tight gas field in the Southern Permian Basin (SPB), East Frisia, Northern Germany. To improve field development, 3D seismic data, wireline and core data were compared to a reservoir analogue in the Panamint Valley, California, United States. Depositional environments of the Permian Upper Rotliegend II included perennial saline lakes, coastal parallel sand belts comprising wet, damp and dry sand flats and aeolian dunes with interdune deposits. Polygonal patterns at different scales were observed on seismic horizon slices in the reservoir intervals and the overlying Zechstein. Outlines of superordinate polygons coincide with interpreted faults. Similar polygonal networks were identified on satellite images of modern dry lakes in the western United States. The mega-crack pattern of ephemeral dry North Panamint Lake, United States, is characterised by variably sized polygons with diameters ranging from kilometres to metres. The evolution and subsurface extent of this polygonal pattern and a potential tectonic link was examined by ground resistivity measurements and surface mapping. Crack development is initiated by shrinking of clays due to changes in water content in the near surface range. For crack evolution the following processes are proposed: Cavities develop in $\sim$1 m (3 ft) depth during a subsurface phase, followed by an upward progression and the evolution of the actual surface cracks with the collapse of the overburden into the existing cavities. Cracks are filled by wind-blown sand and dried-out lake sediments from collapsing crack walls. Following burial, differences in competence between crack-fill and surrounding playa lake sediments provide zones of structural weakness that might channelize stress release and faulting. Ground resistivity measurements confirmed the extent of the cracks to a depth of more than 3 m (9 ft). The maximum horizontal influence area expressed by increased moisture contents on the playa surface is up to 12 m (36 ft). Such fissures systems may develop into weakness zones that serve as fault grain and impact reservoir quality in terms of hydraulic connectivity of reservoir compartments. For the Rotliegend reservoirs, porosity and permeability in the reservoir interval was inverted/ decreased by cementation along migration pathways during diagenesis.

P. Antrett, *Characterization of an Upper Permian Tight Gas Reservoir*,
Springer Theses, DOI: 10.1007/978-3-642-36294-1_1,

Permeability barriers and compartmentalised reservoirs are a potential result of this development.

In addition to the large scale approach analysing seismic data on the regional and reservoir dimension, a workflow that investigates microporosity by combining Scanning Electron Microscopy-Broad Ion Beam (SEM-BIB) and optical microscopy was developed. The SEM-BIB technique displays pores down to nm-scale on polished surfaces ($\sim$area of 1.5 mm^2). The ion beam preparation thus provides unprecedented insights into pore geometries and morphologies of tight rock formations in general. As a result, a relative timing of crystallization was established and multiple phases of clay mineral growth identified. Furthermore, the study focused on pore orientation analysis and stress indicator identification using image analysis as a tool. As a first step, a workflow for an upscaling from the SEM images to thin sections was developed.

For a better understanding of the depositional environment and reservoir rock distribution in the SPB, a sedimentary facies analysis of four cores from the tight gas field in East Frisia was compared to a second study area in northern central Germany. The focus lay on a comparison of the depositional environments and palaeo-geographic positions in the SPB during the Upper Rotliegend II (Upper Permian). In contrast to the tight gas field, which was located at a marginal position of the SPB, the second interest area was situated closer to the great saline lake at the centre of the SPB. While the sedimentation was dominated by dry aeolian sediments in East Frisia, deposition in northern central Germany mainly took place under sub-aquatic conditions. The lake level changes of the Rotliegend lake are expressed as cyclicity in the sediments of the cores.

This study demonstrates that tight gas exploration and production requires multidisciplinary, multiscale approaches beyond standard seismic interpretation workflows to better understand the temporal and spatial evolution of these complex reservoirs.

1.1 Rationale

Tight gas reservoirs (Fig. 1.1) play an increasing role in the global search for hydrocarbon resources, by forming a vast but still underexplored and poorly understood resource base. With conventional hydrocarbon reserves continuously decreasing, the global need for a better understanding of tight gas reservoirs is of considerable economic value. Over the last years, the interest in these specific plays has increased dramatically, particularly in the United States (Fig. 1.2), Canada, India and Northern Africa.

However, there is a lack of a consistent definition of a tight gas reservoir. Law and Curtis [17] introduced a standard which defines low permeability sandstone reservoirs exhibiting a permeability lower than 0.1 mD as tight for the United States. In contrast, the German Society for Petroleum, Coal Science and Technology (DGMK) established the critical value for a tight sandstone at 0.6 mD [22].

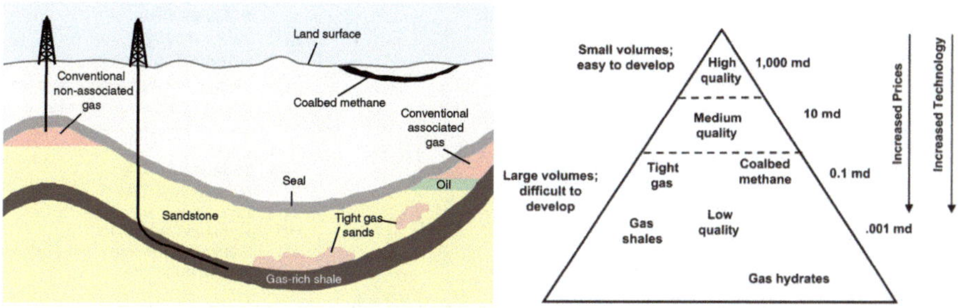

Fig. 1.1 *Left* Schematic geology of natural gas resources (compiled from EIA). *Right* Resource *triangle* for natural gas [15]

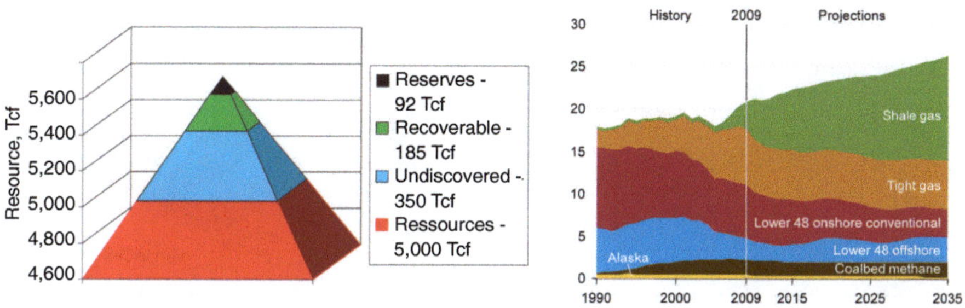

Fig. 1.2 *Left* Resource *triangle* for tight gas in the U.S. [15]. *Right* U.S. natural gas production 1990–2035 in trillion cubic feet per year [32]

Holditch [15] defines a tight gas reservoir as "a reservoir that cannot be produced at economic flow rates nor recover economic volumes of natural gas unless the well is stimulated by a large hydraulic fracture treatment or produced by use of a horizontal wellbore or multilateral wellbores". Despite the scientific and financial challenges, the importance of tight gas reservoirs has been recognised during the last years and great efforts concerning economic exploration and production has been made, especially in the United States. According to Law [16], Haines [14] and Holditch [15], the contingent of natural gas recovered from tight gas reservoirs is significant and economically considerable, ranging between 15 and 20 % (Fig. 1.2).

In the Southern Permian Basin (SPB) in northern Germany (Fig. 1.3), tight gas reservoirs in aeolian sediments represent the main target for the Rotliegend gas exploration. Reserves are estimated to range between 100 and 150 billion cubic metres [3]. Hydrocarbon generation from Westphalian source rocks started during the Early Triassic in deep graben areas. Migration into highly elevated horst blocks was delayed until the Late Cretaceous [26]. Previous studies by Gaupp et al. [7] and Gaupp and Solms [6] proposed that during diagenesis reservoir rocks might have undergone increased degradation of permeability and porosity due to

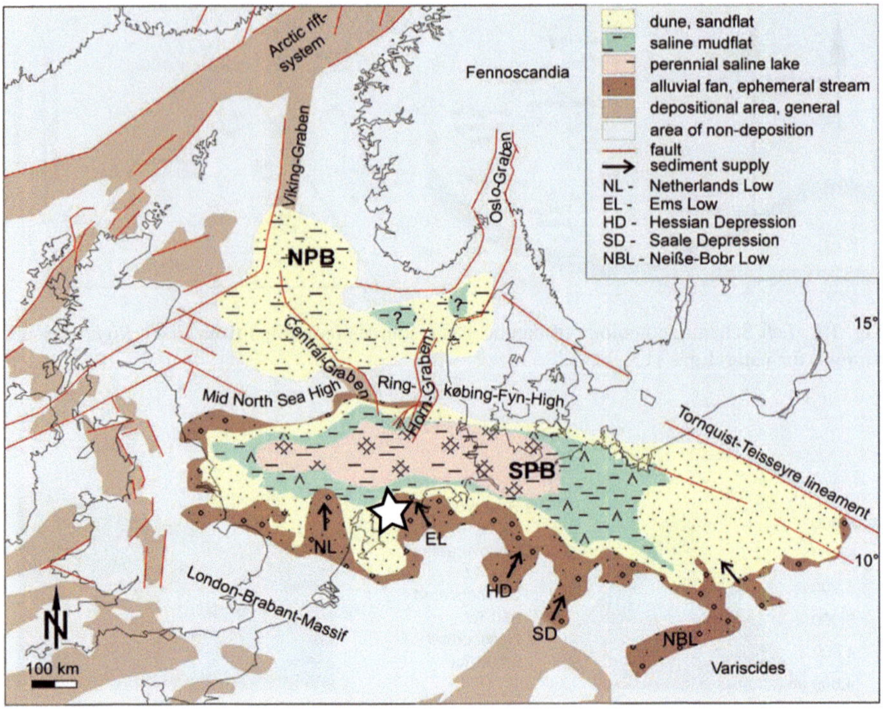

Fig. 1.3 Palaeogeography and general facies distribution of the Upper Rotliegend II Elbe Subgroup (compiled from [13, 29, 37]). *NPB* Northern Permian Basin, *SPB* Southern Permian Basin. The *white star* marks the study area

cross-formation fluid flow and associated enhanced cementation in the vicinity of major faults. Fault-associated cements include illite, chlorite and kaolinite. The processes and timing of this diagenetic effect and its relation to the stress induced fluid pathways are not fully understood at this time.

A possible result of diagenesis and the structural evolution of a play is the development of compartments within the reservoir leading to decreased connectivity and poor reservoir quality. Identification of migration boundaries in reservoirs and unravelling potential controls on their formation plays an important role in today's hydrocarbon exploration. However, it is not always possible to make a precise prediction of the reservoir properties relying solely on seismic data. Although information from cores and wire line logs improve calculations, fast decreasing production rates can often not be satisfactorily explained or predicted. Common solutions involve the drilling of additional expensive wells and the application of hydraulic fracturing techniques. Consequently, a qualitative analysis of tight gas reservoirs requires the establishment of integrated multidisciplinary, multiscale approaches beyond standard seismic interpretation.

1.2 Objectives

This study focuses on a Rotliegend tight gas field in East Frisia in Northern Germany (Fig. 1.3). The main objective lay on the question to which extent seismic data provides information about reservoir properties and reservoir quality. To improve the understanding of early diagenetic processes, development and evolution of hydrocarbon reservoirs in arid climate settings, a modern analogue of the Northern German study area was identified and analysed in the western United States. Results were compared to the information from seismic data and well data of the tight gas field in East Frisia.

In contrast to the large scale approach analysing seismic data on the regional and reservoir scale, a workflow investigating microporosity combining Scanning Electron Microscopy-Broad Ion Beam (SEM-BIB) and optical microscopy was developed in the course of the study.

During the project period of three years the following challenges came up:

1. Evaluate if seismic data provides enough information about reservoir properties/quality of the tight gas field in Northern Germany.
2. Develop an upscaling workflow that quantifies porosity and establishes a link between Scanning Electron Microscopy—Broad Ion Beam (SEM-BIB) and optical microscopy.

1.3 Thesis Outline

The thesis is organised into 7 chapters, which include self-contained information in the main chapters and are summarised in Chap. 7.

In Chap. 2 available data, methodology and tools are introduced.

Chapter 3 deals with the question whether seismic attribute analysis can provide a direct link to reservoir quality in the study area. An approach for the detection of compartmentalised reservoirs in aeolian deposits based on 3D seismic data is introduced. Time-slices are used to delineate potentially useful polygonal boundaries that may help to identify porosity and permeability compartments. By examining amplitude-dependant attributes of horizon slices and surfaces it is possible to map these patterns and take them into consideration for further reservoir analysis. Concerning future field exploration this study provides novel constraints to locate future well sites. The approach presented can be applied in exploration and production in general and to tight gas plays in geologically similar areas worldwide.

In Chap. 4, reasons for the compartmentalisation of hydrocarbon reservoirs are investigated on the example of the Upper Rotliegend II tight gas field in northern Germany that was discussed in Chap. 3. Seismic analysis of the reservoir interval reveals the presence of conspicuous polygonal features, representing fault and fracture systems, which potentially channelise fluid flow and thus may control reservoir quality. To improve the understanding of the origin and evolution of the Rotliegend polygonal systems, the area of the Death Valley National Park, California, United States with the focus on the Panamint Valley region has been chosen as a modern field analogue.

Chapter 5 includes a sedimentary facies analysis of 4 wells situated in northern central Germany. Results are compared to the study area in East Frisia to get a better understanding of the sedimentary facies, palaeogeography and reservoir rock distribution in the Southern Permian Basin during the Upper Rotliegend II.

In Chap. 6, a different approach for the analysis of the tight gas reservoir is chosen. In this chapter macroscopic data, seismic data and well data built the foundation for assumptions about reservoir properties. In this chapter, the SEM-BIB (Scanning Electron Microscopy—Broad Ion Beam) technique is used to display pores down to nm-scale on polished surfaces ($\sim$area of 1.5 mm^2). The ion beam preparation thus provides unprecedented insights into pore geometries and morphologies of tight rock formations in general. In the first phase of the project, selected pores on the polished surface of Rotliegend tight gas samples were investigated. Initially, different phases of mineral growth could be identified and the relative timing of crystallisation was established. Additionally, work focused on pore orientation analysis and stress indicator identification using image analysis as a tool. As a first step, a workflow for an upscaling from the SEM images to thin sections was developed with the intention to extend the routine step wise in future work from thin sections to plugs, and plugs to cores.

Chapter 7 briefly summarises the results and conclusions of this thesis and provides an outlook on future projects.

1.4 The Rotliegend of the Southern Permian Basin in Northern Germany

1.4.1 Tectonic Setting

The east–west trending Southern Permian Basin (SPB) extends from the United Kingdom across the southern North Sea through northern Germany into Poland and the Baltic States. Its width varies between 300 and 600 km (Fig. 1.3). Originally, the basin centre was located in Northern Germany. Due to the ongoing basin development the depositional area enlarged towards the east and west through time [29]. At its maximum extent the basin covered an area of 429.000 km^2.

The thermal subsidence that followed the Lower Rotliegend bimodal magmatism in the SPB throughout the Upper Rotliegend I was superposed by a new phase of extensional tectonics starting with the Upper Rotliegend II [1, 34]. The re-orientations of plate boundary forces [25, 28] resulted in the development of a widespread erosional unconformity underlying the Upper Rotliegend II strata in

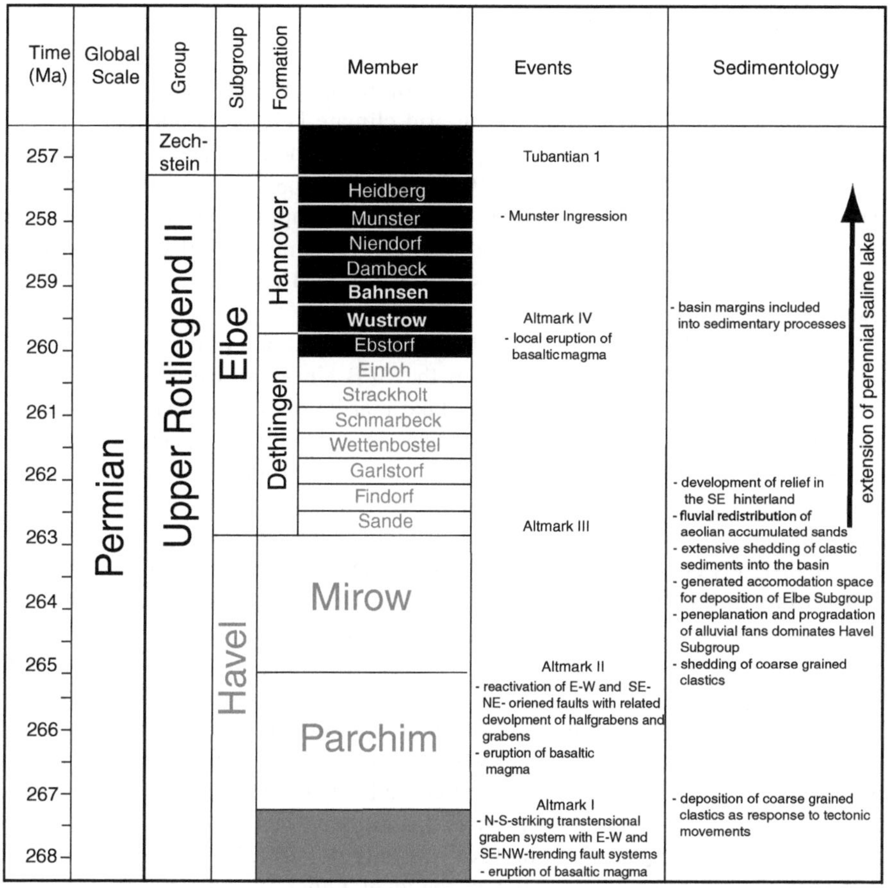

Fig. 1.4 Stratigraphic chart of the Upper Permian of Northern Germany (modified from [23, 18]; timescale according to [4]). Interval of seismic interpretation highlighted in *black*. Reservoir interval in *bold*

central and northern Germany [8, 25]. At least two of the four Altmark tectonic events during the deposition of the Havel/Elbe Subgroup were associated with basaltic volcanism ([9]; Fig. 1.4). Volcanic deposits could be observed locally, situated beneath the reservoirs. The development of transtensive and transpressive structures, frequently accompanied by soft sediment deformation were the result of the Altmark tectonic pulses. Successive rifting events that ultimately caused the opening of the Tethys and the Arctic-North Atlantic Oceans went along with localised thermal uplift of the crust elsewhere and resulted in phases of enhanced heat flow, erosion and the development of pre-Zechstein unconformities in regions bordering the Central European Basin [29].

1.4.2 Depositional Setting

The Upper Rotliegend II of the SPB is dominated by siliciclastics and minor evaporites deposited under arid to semi-arid climate [11, 12]. Four facies associations were identified to reflect the deposition in ephemeral fluvial (wadi), aeolian, sabkha and lacustrine environments [10, 18, 31]; Fig. 1.3). According to Gast et al. [5] and Rieke [24] a huge amount of sediment was supplied by aeolian transport through strong easterly winds. In addition, the remaining relief of the Variscan Orogen and local highs towards the south of the SPB sourced ephemeral fluvial systems that drained into the SPB.

The existence of a perennial saline lake in the central part of the SPB provided a major control on base level changes in the SPB since the Dehlingen Subgroup (Fig. 1.4). Due to the low relief of the lake margin areas, even small lake level fluctuations favoured considerable changes in depositional facies. During the deposition of the Elbe Subgroup (Fig. 1.4), two-third-order tectonic cycles, Altmark III and IV, consistent of 14 fourth-order transgressive and regressive cycles, each with a duration of ~400 ka [4] record these lake level changes. During lake level highstands claystones were deposited, whereas halite formation was dominant during times of lake level lowstands and enhanced isolation. Lake sedimentation was accompanied by coastal parallel accumulation of sand belts, comprising successions of wet, damp and dry sand flats and aeolian dunes with interdune deposits. The prevailing easterly and north-easterly winds played an important role in the accumulation of aeolian deposits, respectively sand deflation in some areas [12]. Characteristic events of short-term marine ingressions were proposed by Legler [18] on the basis of marine fauna (liebea reichei). According to Van Wees et al. [34], such short-term but widespread ingression of marine waters into the Upper Rotliegend II saline lake was only possible, because the basin floor of the SPB had already subsided below sea level. Other explanations for the short-term marine events consider heavy storms or tsunamis in combination with sea level highstands in the Arctic rift system [27]. The Munster ingression represents an example of an ingression forced by tectonic movements as it did not occur contemporaneous to a sea level highstand [18]. In any case, the basin did not achieve a fully marine character until the onset of the major Zechstein transgression.

With the Zechstein transgression, the dominantly terrestrial sedimentation of the Rotliegend ceased and successive marine sedimentation cycles started [2, 30]. Zechstein deposits reach thicknesses of ~1,500 m in the SPB [29].

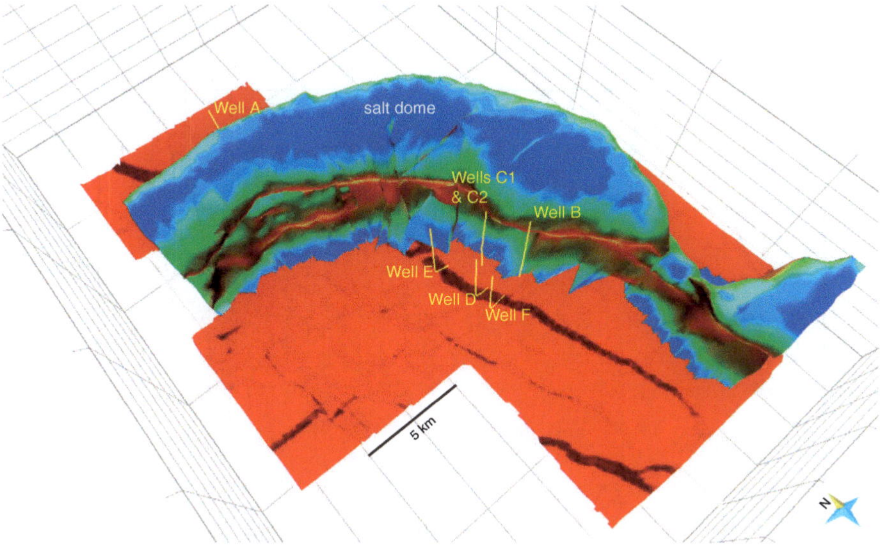

Fig. 1.5 Overview of the interest area on 3D seismic data. Salt dome (*blue-green*), Top Rotliegend surface (*red*) and wells A–E (*yellow*) displayed

1.5 Study Area

The Rotliegend graben/halfgraben system (Fig. 1.5) is overlain by a prominent unconformity that developed during early Buntsandstein times (Fig. 1.6). Compared to the Permian, the orientation of stress and deformation changed [33]. The early Triassic was characterised by the onset of thin-skinned extension [19]. The main phase of salt movement started in the middle to late Triassic and continued throughout the Cretaceous [20] accompanied by salt withdrawal and downbuilding (Fig. 1.6). As a result of salt diapirism, thicknesses of Zechstein deposits vary (Fig. 1.6). In front of major faults, wedge-like structures can be observed where salt was trapped during the removal process [33]. A striking unconformity developed at the base of the Cretaceous [35]. Furthermore salt rim synclines, adjacent to the salt dome suggest the presence of a topographic relief as a result of salt piercing the surface [21, 36]. Salt deformation still affected the sedimentation during the Tertiary (Fig. 1.6).

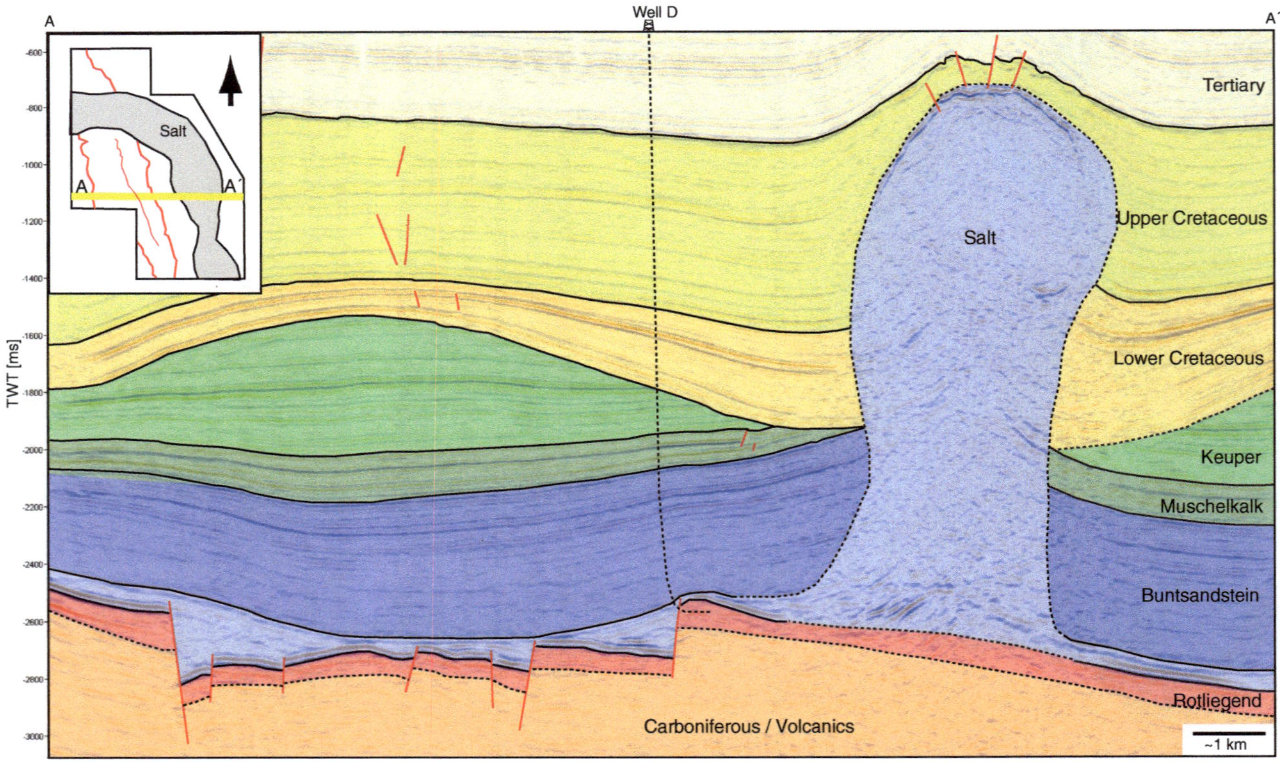

Fig. 1.6 E–W oriented cross section with Well D sketched and major faults and reflectors interpreted. Rotliegend graben/halfgraben system overlain by unconformity in early Buntsandstein times. Steady sedimentation until Base Cretaceous unconformity where high salt movement causes downbuilding and the development of salt rim synclines. Base Tertiary affected by salt movement [33]

References

1. Bachmann GH, Hoffmann N (1997) Development of the Rotliegend Basin in Northern Germany. Geol Jahrb D 103:9–31
2. Becker F, Bechstädt T (2006) Sequence stratigraphy of a carbonate-evaporite succession (Zechstein 1, Hessian Basin, Germany). Sedimentology 53:1083–1120
3. BGR (2009) In: Cramer B, Andruleit H (eds) Energierohstoffe 2009: Reserven, Ressourcen, Verfügbarkeit—Erdöl, Erdgas, Kohle, Kernbrennstoffe, Geothermische Energie
4. Gast R (1995) Sequenzstratigraphie. In: Plein E (ed) Stratigraphie von Deutschland I; Norddeutsches Rotliegendbecken: Rotliegend-Monographie Teil II, vol 183. Courier Forschungs-Institut Senckenberg, Frankfurt a. M., pp 47–54
5. Gast R, Pasternak M, Piske J, Rasch H-J (1998) Das Rotliegend im nordostdeutschen Raum: Regionale Übersicht, Stratigraphie, Fazies und Diagenese. Geol Jahrb A 149:59–79
6. Gaupp R, Solms M (2005) Sedimentological and Petrological investigations. In: Paleo Oil and Gasfields in the Rotliegend of the North German Basin: effects upon Hydrocarbon Reservoir Quality (Paläo-Öl- und Gasfelder im Rotliegenden des Norddeutschen Beckens: Wirkungen der KW-Migration auf die Speicherqualitäts-Entwicklung.) DGMK Research Report, 593–598: Tight Gas Reservoirs—Natural Gas for the Future 1–1, 1–44
7. Gaupp R, Matter A, Platt J, Ramseyer K, Walzebuck JP (1993) Diagenesis and fluid evolution in deeply buried Permian (Rotliegend) gas reservoirs, Northwest Germany. AAPG Bull 77(7):1111–1128
8. Gebhardt U, Plein E (1995) Neue Gliederung. In: Plein E (ed) Norddeutsches Rotliegendbecken, Rotliegend-Monographie Teil II, Courier Forschungs-Institut Senckenberg, Frankfurt a. M. 183:18–24
9. Gebhardt U, Schneider J, Hoffmann N (1991) Modelle zur Stratigraphie und Beckenentwicklung im Rotliegenden der Norddeutschen Senke. Geol Jahrb A 127:405–427
10. George GT, Berry JK (1993) A new palaeogeographic and depositional model for the Upper Rotliegend of the UK Sector of the Southern North Sea. In: North CP, Prosser DJ (eds) Characterization of Fluvial and Aeolian reservoirs, vol 73. Geological Society, London, Special Publications, pp 291–319
11. Glennie KW (1972) Permian Rotliegendes of Northwest Europe interpreted in light of modern desert sedimentation studies. AAPG Bull 56(6):1048–1071
12. Glennie KW (1983) Early Permian (Rotliegendes) palaeowinds of the North Sea. Sed Geol 34:245–265
13. Glennie KW, Higham J, Stemmerik L (2003) Permian. In: The millennium atlas; petroleum geology of the central and northern North Sea. Geological Society of London, London, pp 91–103
14. Haines L (ed) (2005) Tight Gas. Supplement to Oil and Gas investor, March
15. Holditch SA (2006) Tight gas sands. J Pet Technol 86–93
16. Law BE (2002) Basin-centered gas systems. AAPG Bull 86(11):1891–1919
17. Law BE, Curtis JB (2002) Introduction to unconventional petroleum systems. AAPG Bull 86(11):1851–1852
18. Legler B (2005) Faziesentwicklung im Südlichen Permbecken in Abhängigkeit von Tektonik, eustatischen Meeresspiegelschwankungen des Proto-Atlantiks und Klimavariabilität (Oberrotliegend, Nordwesteuropa). Schriftenreihe der Deutschen Gesellschaft für Geowissenschaften, 47
19. Lohr T, Krawczyk CM, Tanner DC, Samiee R, Endres H, Thierer PO, Oncken O, Trappe H, Bachmann R, Kukla PA (2008) Prediction of subseismic faults and fractures: integration of three-dimensional seismic data, three-dimensional retrodeformation, and well data on an example of deformation around an inverted fault. AAPG Bull 92(4):473–485
20. Mohr M, Kukla PA, Urai JL, Bresser G (2005) Multiphase salt tectonic evolution in NW Germany; seismic interpretation and retro-deformation. Int J Earth Sci 94(5–6):917–940

21. Mohr M, Warren JK, Kukla PA, Urai JL, Irmen A (2007) Subsurface seismic record of salt glaciers in an extensional intracontinental setting (Late Triassic of northwestern Germany). Geology 35:96–966
22. Naik GC (2005) Tight Gas Reservoirs: an unconventional natural energy source for the future
23. Plein E (1995) Stratigraphie von Deutschland I: Norddeutsches Rotliegendbecken, Rotliegend-Monographie Teil II, Courier Forschungs-Institut Senckenberg, Frankfurt a. M., 183
24. Rieke H (2001) Sedimentologie, Faziesarchitektur und Faziesentwicklung des kontinentalen Rotliegenden im Nordostdeutschen Becken (NEDB). Ph.D. thesis, Potsdam University, Germany
25. Schneider J, Gebhardt U, Gaitzsch B (1995a) Fossilführung und Stratigraphie. In: Plein E (ed) Norddeutsches Rotliegendbecken, Rotliegend-Monographie Teil II. Courier Forschungs-Institut Senckenberg, Frankfurt a. M. 183:25–39
26. Schwarzer D, Littke R (2007) Petroleum generation and migration in the 'Tight Gas' area of the Germany Rotliegend natural gas play: a basin modelling study. Pet Geosci 13:37–62
27. Stemmerik L (2001) Sequence stratigraphy of a low productivity carbonate platform succession: the Upper Permian Wegener Halvø Formation, Karstryggen Area, East Greenland. Sedimentology 48:79–97
28. Stollhofen H (2007) Postvulkanische Rotliegend-Schwemmfächersysteme am Hunsrück-Südrand, Saar-Nahe-Becken, SW-Deutschland. Jahresberichte und Mitteilungen des Oberrheinischen Geologischen Vereins 89:285–306
29. Stollhofen H, Bachmann GH, Barnasch J, Bayer U, Beutler G, Franz M, Kästner M, Legler B, Mutterlose J, Radies D (2008) Upper Rotliegend to Early Cretaceous basin development. In: Littke R, Bayer U, Gajewski D, Nelskamp S (eds) Dynamics of complex intracontinental basins. The Central European Basin system. Springer, Berlin, Heidelberg, 181–210
30. Strohmenger C, Voigt E, Zimdars J (1996) Sequence stratigraphy and cyclic development of Basal Zechstein carbonate-evaporite deposits with emphasis on Zechstein 2 off-platform carbonates (Upper Permian, Northeast Germany). Sed Geol 102:33–54
31. Strömbäck AC, Howell JA (2002) Predicting distribution of remobilized aeolian facies using sub-surface data: the Weissliegend of the UK Southern North Sea. Pet Geosci 8:237–249
32. U.S. Energy Information Administration (2011) Annual Energy Outlook 2011 with Projections to 2035. April 2011. DOE/EIA-0383
33. Vackiner A, Antrett P, Stollhofen H, Back S, Kukla PA, Bärle C (2011) Syndepositional Tectonic Controls and Palaeo-Topography of a Pemrian tight gas reservoir in NW Germany. J Pet Geol 34(4):411–428
34. Van Wees J-D, Stephenson RA, Ziegler PA, Bayer U, McCann T, Dadlez R, Gaupp R, Narkiewicz M, Bitzer F, Scheck M (2000) On the origin of the Southern Permian Basin, Central Europe. Mar Pet Geol 17:43–59
35. Voigt T, Reicherter K, von Eynatten H, Littke R, Voigt S, Kley J (2008) Sedimentation during Basin Inversion. In: Littke R, Bayer U, Gajewski D, Nelskamp S (eds) Dynamics of complex intracontinental basins. The Central European Basin System. Springer, Berlin, Heidelberg, pp 211–232
36. Warren JK (2008) Salt as sediment in the Central European Basin system as seen from a deep time perspective. In: Dynamics of complex intracontinental basins; the Central European Basin system. Springer, Berlin, Federal Republic of Germany, pp 249–276
37. Ziegler PA (1982) Geological Atlas of Western and Central Europe. Shell Internat. Petrol Maatsch, The Hague

Chapter 2
Data and Methodology

This chapter is based on multidisciplinary field data. A brief overview of the available data sets and the applied methodology is provided in this chapter. A description in greater detail is given in the specific chapters.

The 3D seismic survey analysed in Chaps. 3 and 4 is located in East Frisia and is the result of the merging of three individual data volumes acquired from 1995 to 2001. It covers an area of $\sim 293\ \mathrm{km}^2$. In addition to the data set in the time domain, a depth converted volume was available. Analysis was mainly carried out in time domain due to the fact, that subtle information was lost during later stages of processing, filtering and smoothing. Volumes of amplitude dependent and amplitude independent seismic attributes were processed using standard seismic interpretation software packages.

Wireline logs from seven wells and core material from four of these wells were available from the East Frisia study area. Formation Micro Imaging/Formation Micro Scanner (FMS/FMI) data are available from one of the wells. Core analysis discussed in Chaps. 3 and 4 was carried out macroscopically. Differentiation between sedimentary environments is based on clay content and internal structures such as grain size distribution and internal geometries.

Ground resistivity measurements in the Panamint Valley in California, United States (Chap. 4) were carried out using a "Lippmann 4-point light" geoelectrical instrument. Data was processed with the Res2DInv software. Collected sediment samples were investigated by x-ray diffractometry analysis (XRD), glycol dehydration and optical microscopy. In addition to the field work, satellite images provided by the Google Earth Software [3] were taken into consideration. Earthquake data was available from the United States Geological Survey Northern California Seismic Network (USGS NCSN) Catalog (http://www.ncedc.org/).

Core material from four wells in northern central Germany (Chap. 5) was analysed macroscopically with the focus on the interpretation of the depositional sedimentary environment. Similar to the cores discussed in Chaps. 3 and 4, differentiation between sedimentary environments is based on differences in grain sizes and internal geometries.

P. Antrett, *Characterization of an Upper Permian Tight Gas Reservoir*,
Springer Theses, DOI: 10.1007/978-3-642-36294-1_2,

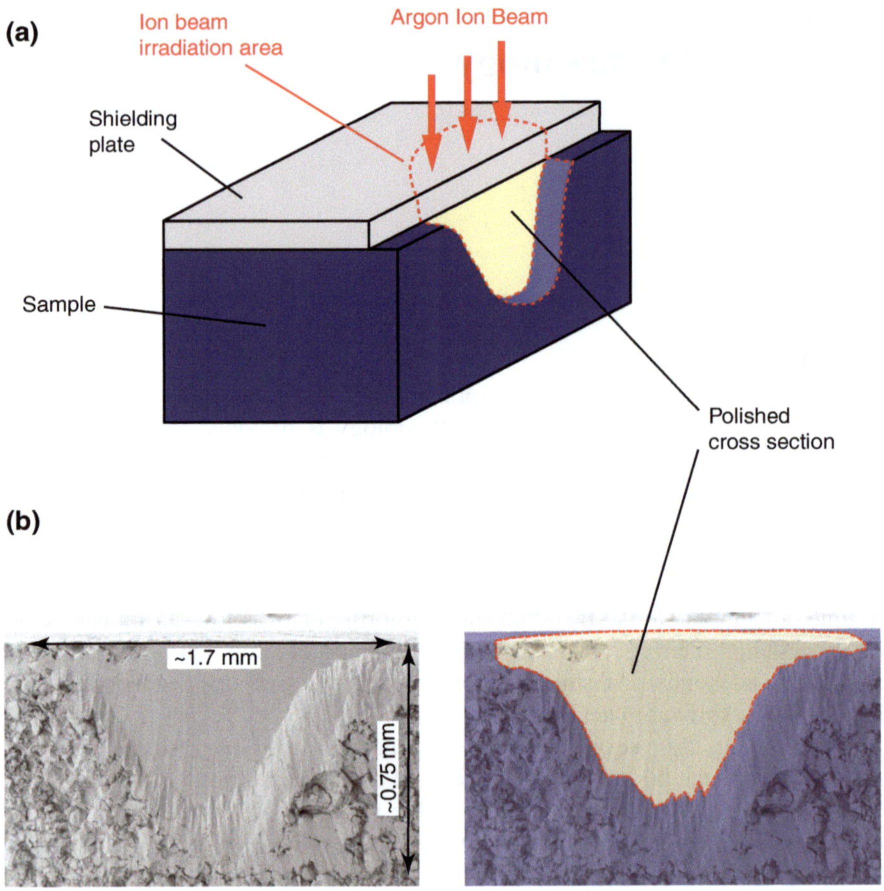

Fig. 2.1 The principle of BIB cross-sectioning. **a** The ion beam irradiates the edge of sample un-masked by the shielding plate to create high quality polished cross-sections suitable for SEM imaging. **b** Overview of a typical cross section performed by BIB cross sectioning (A: modified after Desbois et al. [3]

As described in Chap. 6, SEM-BIB samples were polished using the JEOL SM-09010 [2]; Fig. 2.1) argon ion beam cross-section polisher and investigated with Zeiss-SUPRA 55 SEM microscope with secondary electron detector (SE), back-scattered electrons (BSE) and energy-dispersive X-ray (EDX) (EDAX-Apollo10 SDD) detectors [1]. Additionally, optical microscopy analysis was performed using a Leica DM 4500 polarisation microscope. SEM-BIB and thin section images were evaluated with the focus on mineral and porosity percentages with the JMicroVision software Version 1.2.7 [4].

References

1. Desbois G, Urai JL, Kukla PA, Konstanty J, Baerle C (2011) High-resolution 3D fabric and porosity model in a tight gas sandstone reservoir: a new approach to investigate microstructures from mm- to nm-scale combining argon beam cross-sectioning and SEM imaging. J Pet Sci Eng. doi:10.1016/j.petrol.2011.06.004
2. Erdmann N, Campbell R, Asahina S (2006) Precise SEM cross-section polishing via Argon beam milling. Microscopy Today, 22–25 May 2006
3. Google Inc. (2011) Google Earth (Version 5.2.1.1588) [Software]. http://www.google.com/earth/index.html
4. Roduit N (2011) JMicroVision: image analysis toolbox for measuring and quantifying components of high-definition images. Version 1.2.7. http://www.jmicrovision.com, accessed May 2011

Chapter 3
Seismic Attribute Analysis for Detection of Highly Compartmentalised Reservoirs

3.1 Introduction

Tight gas reservoirs play an increasing role in the global search for hydrocarbon resources, by forming a vast, but still underexplored and poorly understood resource. With conventional hydrocarbon reserves continuously decreasing, the global need for a better understanding of tight gas reservoirs is of considerable economic value. Gas reservoirs in aeolian sediments represent the main target for the Rotliegend gas exploration. In the Southern Permian Basin (SPB) in northern Germany, hydrocarbon generation from Westphalian source rocks started during the Early Triassic in deep graben areas. Migration into highly elevated horst blocks was delayed until the Late Cretaceous [38]. Previous studies by Gaupp et al. [11] and Gaupp and Solms [10] proposed that during diagenesis reservoir rocks might have undergone increased degradation of permeability and porosity due to cross-formation fluid flow and associated enhanced cementation in the vicinity of major faults. Fault-associated cements include illite and kaolinite. The processes and timing of this diagenetic effect and its relation to the stress induced fluid pathways are not fully understood at this time.

In this chapter, an approach for detecting compartmentalised reservoirs in aeolian deposits based on 3D seismic data is provided. Time-slices are used to delineate polygonal boundaries that may help to identify potential porosity and permeability compartments. By examining amplitude-dependant attributes of horizon slices and surfaces, these patterns can be mapped and taken into consideration for further reservoir analysis. Concerning future field exploration this study provides novel constraints to locate future well sites. The approach presented can be applied in exploration and production in general and to tight gas plays in geologically similar areas worldwide.

P. Antrett, *Characterization of an Upper Permian Tight Gas Reservoir*,
Springer Theses, DOI: 10.1007/978-3-642-36294-1_3,

3.2 Geological Framework

3.2.1 Tectonic Setting

The east–west trending SPB extends from the United Kingdom across the southern North Sea through northern Germany into Poland and the Baltic States. Its width varies between 300 and 600 km (Fig. 3.1). Originally, the basin centre was located in Northern Germany, but due to the ongoing basin development, the depositional area enlarged towards the east and west through time [40]. At its maximum extent the basin covered an area of 429.000 km^2.

The thermal subsidence that followed the Lower Rotliegend bimodal magmatism in the SPB throughout the Upper Rotliegend I was superposed by a new phase of extensional tectonics starting with the Upper Rotliegend II ([1, 46]; Fig. 3.2). The re-orientations of plate boundary forces [37, 40] resulted in the development of a widespread erosional unconformity underlying the Upper Rotliegend II strata in central and northern Germany [13, 37]. At least two of the four Altmark tectonic events during the deposition of the Havel/Elbe Subgroup were associated with basaltic volcanism ([14]; Fig. 3.2). Volcanic deposits could be observed locally, situated beneath the reservoirs. The development of transtensive and transpressive structures, frequently accompanied by soft sediment deformation were the result of the Altmark tectonic pulses. Successive rifting events that ultimately caused the opening of the Tethys and the Arctic-North Atlantic Oceans went along with

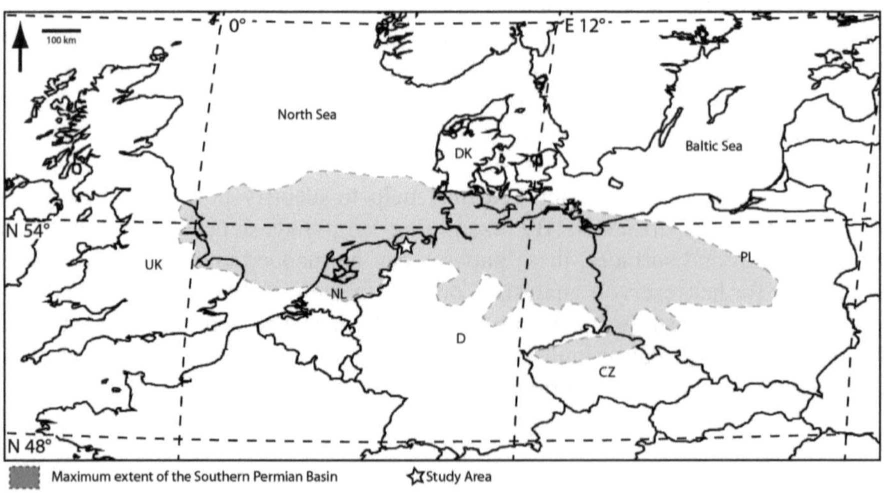

Fig. 3.1 Map of Western Europe with the maximum extent of the SPB during the Upper Rotliegend II (modified from [49]). *Star* marks the study area. *CZ* Czech Republic, *D* Germany, *DK* Denmark, *PL* Poland, *UK* United Kingdom

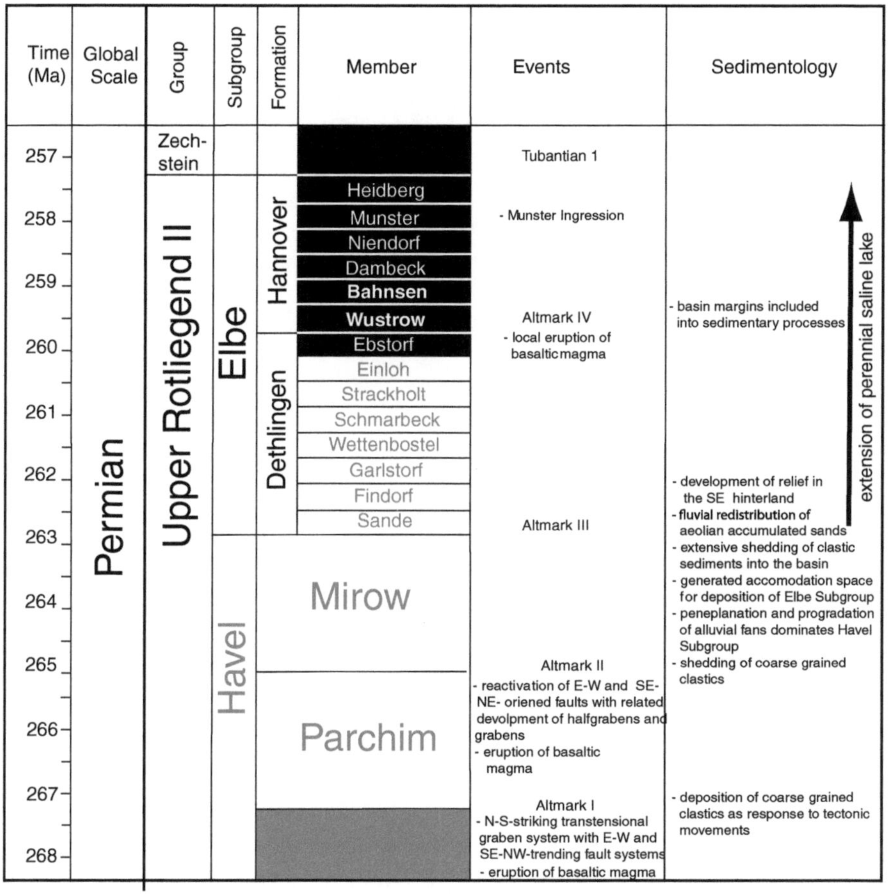

Fig. 3.2 Stratigraphic chart of the Upper Permian in northern Germany (modified from [24, 31]; timescale according to [7]). Interval of seismic interpretation highlighted in *black*. Reservoir interval in *bold*

localized thermal uplift of the crust elsewhere and resulted in phases of enhanced heat flow, erosion and the development of pre-Zechstein unconformities in regions bordering the Central European Basin [40].

The E–W seismic section displayed in Fig. 3.3 provides an overview of the structural framework and the major stratigraphic units occurring in the study area. The N–S trending Rotliegend graben/halfgraben system is overlain by a prominent unconformity that developed during early Buntsandstein times (Fig. 3.3). Compared to the Permian, the orientation of stress and deformation changed [43]. The early Triassic was characterized by the onset of thin-skinned extension [25]. The main phase of salt movement started in the middle to late Triassic and continued throughout the Cretaceous [27] accompanied by salt withdrawal and downbuilding (Fig. 3.3). As a result of salt diapirism, thicknesses of Zechstein deposits vary

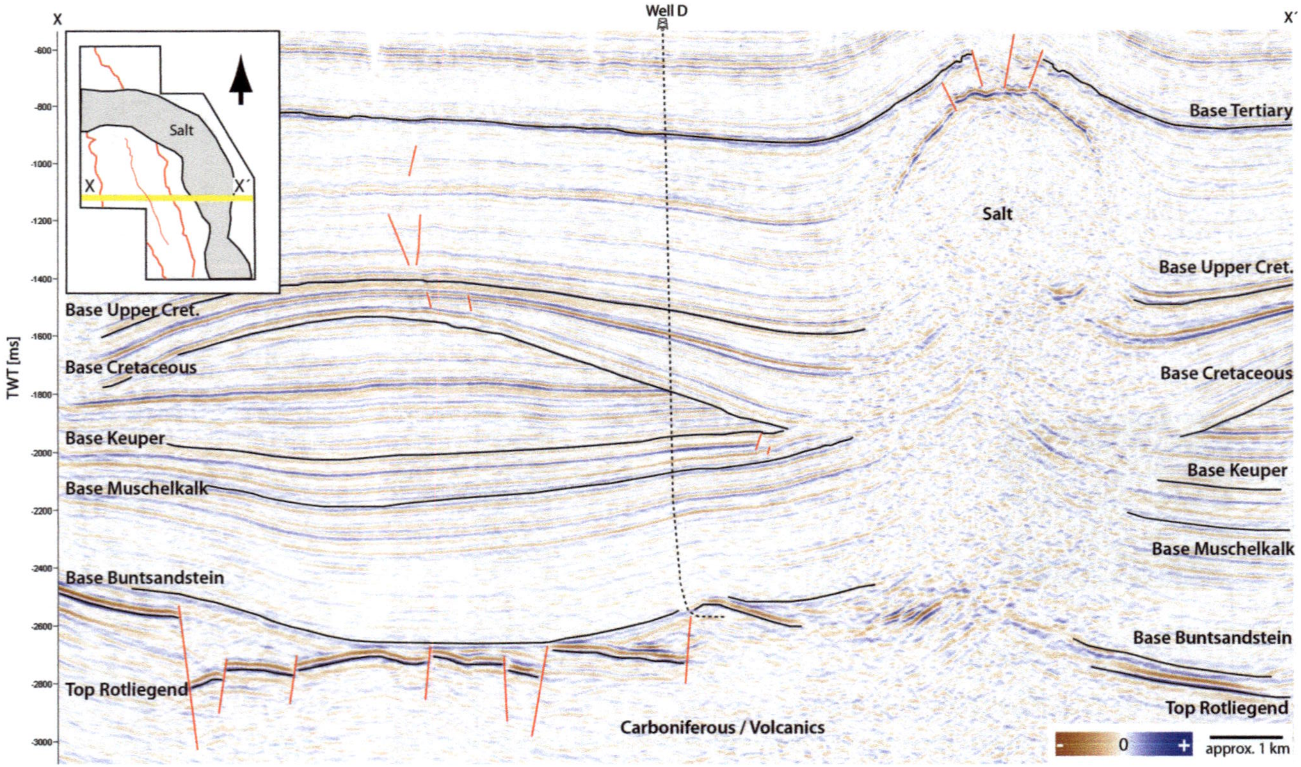

Fig. 3.3 E–W oriented cross section with Well D sketched and major faults and reflectors interpreted. Rotliegend graben/halfgraben system overlain by early Buntsandstein unconformity. Steady sedimentation until Base Cretaceous unconformity where strong salt movement causes downbuilding and the development of salt rim synclines. Base tertiary affected by salt movement [44]

(Fig. 3.3). In front of major faults, wedge-like structures can be observed where salt remained in place during the removal process [43]. A striking unconformity developed at the base of the Cretaceous [47]. Furthermore, salt rim synclines adjacent to the salt dome suggest the presence of a topographic relief as a result of salt piercing the surface [28, 48]. Salt deformation still affected deposition during the Tertiary (Fig. 3.3).

3.2.2 Depositional Setting

The Upper Rotliegend II of the SPB is dominated by siliciclastics and minor evaporites deposited under arid to semi-arid climate [16, 17]. Four facies associations were identified to reflect the deposition in ephemeral fluvial (wadi), aeolian, sabkha and lacustrine environments [15, 24, 42]. According to Gast et al. [8] and Rieke [34] a huge amount of sediment was supplied by aeolian transport through strong easterly winds. In addition, the remaining relief of the Variscan Orogen and local highs towards the south of the SPB sourced ephemeral fluvial systems that drained into the SPB.

The existence of a perennial saline lake in the central part of the SPB provided a major control on base level changes in the SPB since the Dethlingen Subgroup (Fig. 3.2). Due to the low relief of the lake margin areas, even small lake level fluctuations favoured considerable changes in depositional facies. During the deposition of the Elbe Subgroup two third-order tectonic cycles, Altmark III and IV, consistent of 14 fourth-order transgressive and regressive cycles, each with a duration of $\sim$400 ka [7] record these lake level changes. During lake level highstands claystones were deposited, whereas halite formation was dominant during times of lake level lowstands and enhanced isolation. Lake sedimentation was accompanied by coastal-parallel accumulation of sand belts, comprising successions of wet, damp and dry sandflats and aeolian dunes with interdune deposits. The prevailing easterly and north-easterly winds played an important role in the accumulation of aeolian deposits, respectively in sand deflation in some areas [17]. Characteristic events of short-term marine ingressions were proposed by Legler [24] on the basis of marine fauna (liebea reichei). According to Van Wees et al. [45], such short-term but widespread ingression of marine waters into the Upper Rotliegend II saline lake was only possible, because the basin floor of the SPB had already subsided below sea level. Other explanations for the short-term marine events consider heavy storms or tsunamis in combination with sea level highstands in the Arctic rift system [38]. The Munster ingression is documented as an example of a tectonic induced ingression due to the fact that it did not occur contemporaneous to a sea level highstand [24]. In any case, the basin did not achieve a fully marine character until the onset of the major Zechstein transgression.

With the Zechstein transgression, the dominantly terrestrial sedimentation of the Rotliegend ceased and successive marine sedimentation cycles started [2, 41]. Zechstein deposits reach thicknesses of ~1,500 m in the SPB [40].

3.3 Data and Methods

The analysis is based on multidisciplinary field data incorporating a 3D seismic reflection survey (Fig. 3.4), wireline logs from seven wells and sedimentological logging of cores from four wells. The pre-stack time migrated seismic survey is the result of merging three surveys acquired in 1995, 1996 and 2001 that cover a total of 293 km^2. After the merge, the trace length is 5 s and the sample interval 4 ms. The frequency of the data varies between 14 and 70 Hz. The bin size is 25 × 25 m. In this study, all seismic data are displayed in SEG normal polarity (European polarity), in which a positive amplitude represents a decrease in impedance and a negative amplitude an impedance increase [3].

In the investigated study interval, every reflector was interpreted, followed by the generation of series of horizon slices. Cubes of amplitude-dependent attributes (variance, ant tracking, relative acoustic impedance) and amplitude independent attributes (3D curvature) were processed to identify and display subseismic features that may influence reservoir properties. Colour schemes were customized to emphasize small variations of the applied seismic attribute. The resolution of the standard colour bar was extended by inserting additional colours in specific intervals of interest (see Figs. 3.5b and 3.7).

The variance attribute was used to isolate edges (discontinuities in the horizontal continuity of the amplitude) from the input dataset. The processed variance cube then served as input for the calculation of the ant tracking cube using a filter size of 3 for the inline and crossline ranges and a (mild) vertical smoothing of 15 samples.

Ant tracking was used to extract faults from the variance cube. The following parameters were of concern for ant tracking: (1) The number of voxels chosen for the initial boundary, controlling tracking detail; (2) the ant tracking deviations controlling the number of voxels investigated on either side of its trace; (3) the ant step size defining the ant advances for each increment of its search. Resolution decreases with increasing volume size; (4) the number of illegal steps that defined how far an ant searched without reaching an edge; (5) the number of legal steps required after an illegal step; and finally (6) the stop criteria controlling the termination an ant advance after a certain percentage of illegal steps (between 0 and 50 %). Two different variations of presets were applied for the processing of the ant tracking attribute cubes. Firstly, the initial ant boundary was set at seven voxels, an ant track deviation of two voxels and the ant step size of three voxels (passive ant tracking). The calculated cube was then processed a second time with the following presets: an initial ant boundary of five voxels, the ant track deviation of two voxels and an ant step size of three voxels (aggressive ant tracking). The stopping criteria for the first ant tracking was set at 5 % with one illegal step

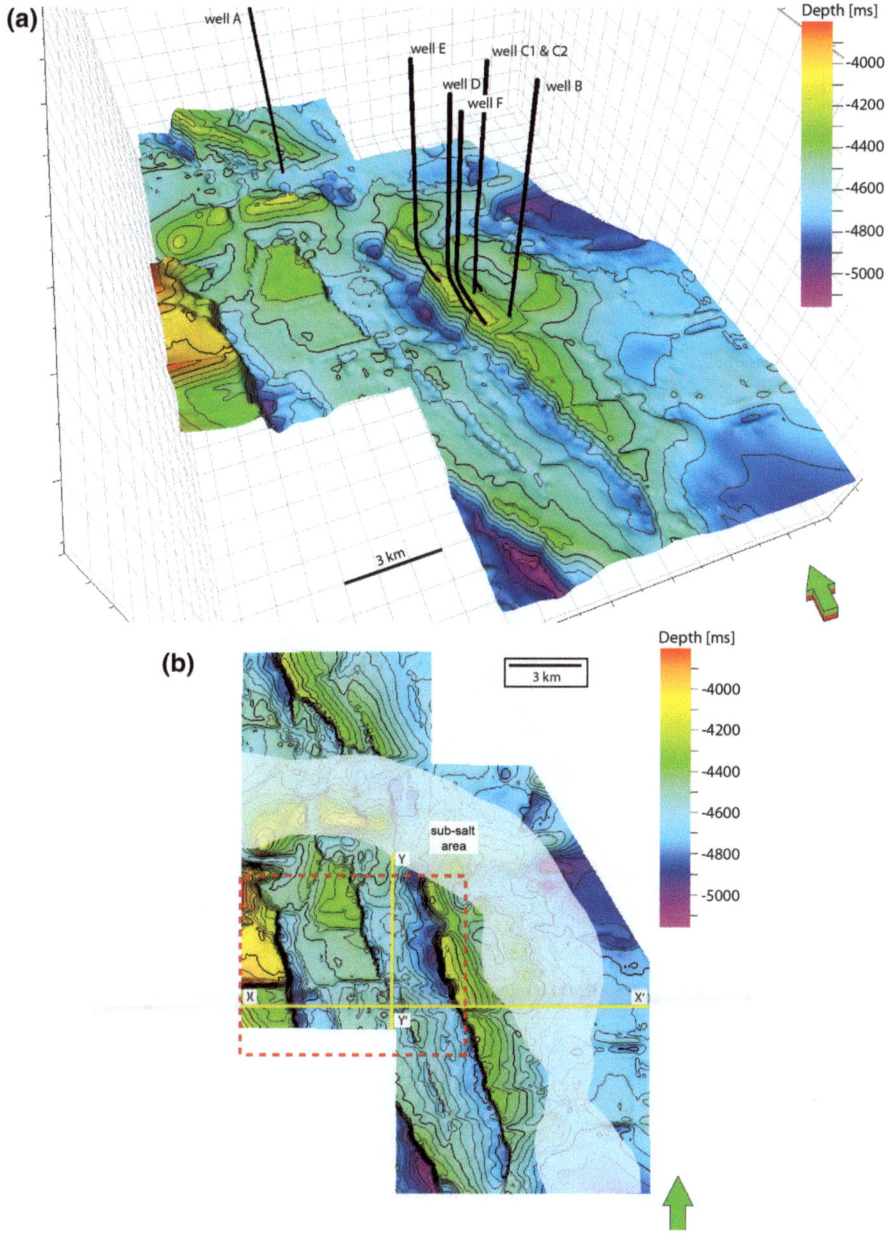

Fig. 3.4 *Upper image* 3D overview of the Top Rotliegend surface including the wells; *Lower image* 2D map of the Top Rotliegend surface with profiles X–X′ (Fig. 3.3), Y–Y′ (Fig. 3.5a). The *red* framed area is shown in detail in Fig. 3.7. The area located under the salt dome is sketched in *white*

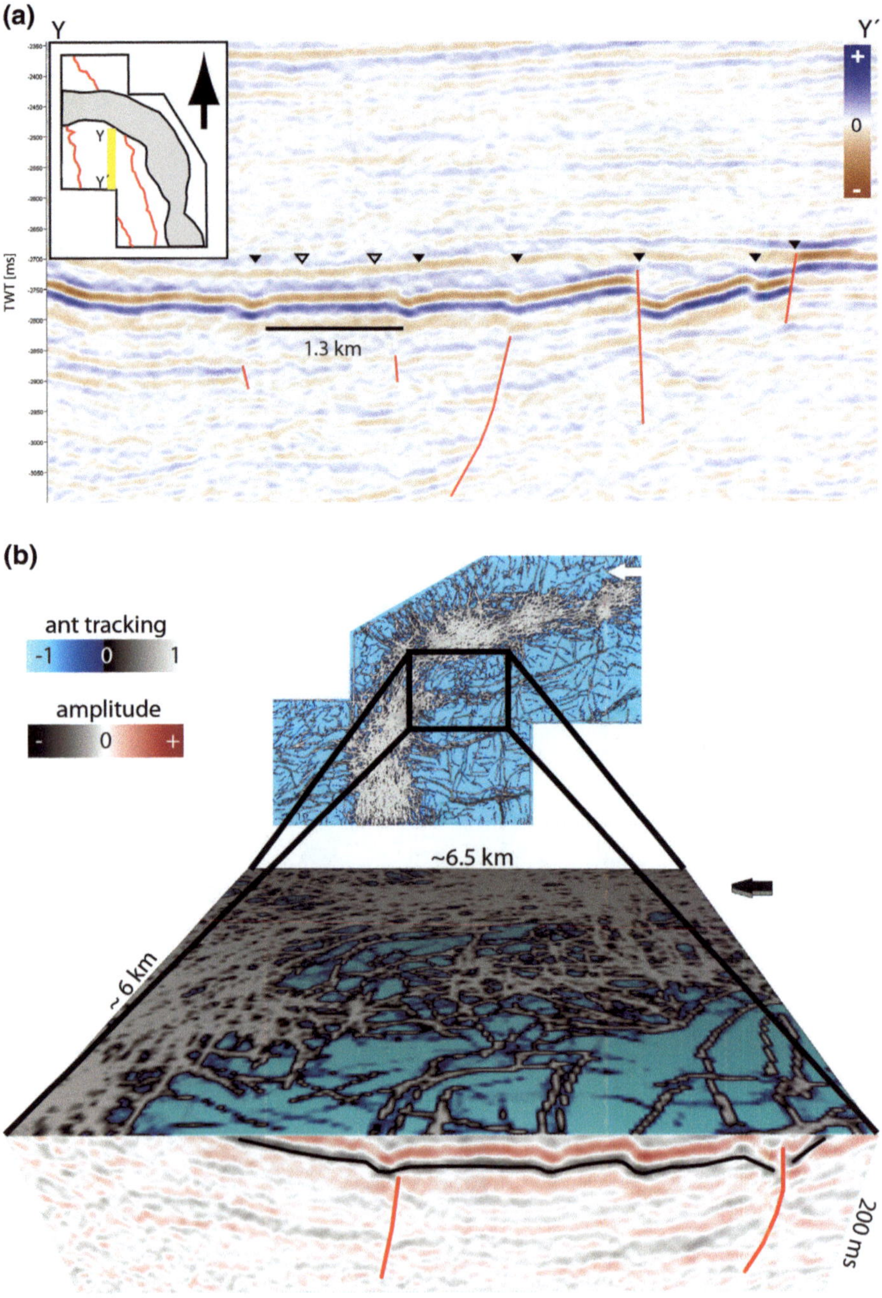
(a)
Y
Y′
+
0
-
TWT [ms]
1.3 km
(b)
ant tracking
-1
0
1
amplitude
-
0
+
~6.5 km
~6 km
200 ms

◀**Fig. 3.5** **a** Section Y-Y′. Top Rotliegend horizon and faults interpreted. *Black arrows* indicate locations of major outlines (see Fig. 3.6), mostly associated with faults on seismic surfaces. *White arrows* mark positions of sub-ordinate outlines with no offset or amplitude variation. **b** Polygonal pattern on cropped seismic volume of an ant tracking z-slice in the foreground. Effect of pattern displayed in vertical seismic line: negative relief of the reflectors not in all cases accompanied by faults. Sometimes no change in amplitude occurs at all. Top Rotliegend horizon sketched in *black*, faults in *red*

allowed and three legal steps required. For the second process, the stop criteria was 10 % with two illegal steps allowed and two legal steps required.

Analysis of the relative acoustic impedance attribute was primarily used for the identification of unconformity surfaces and discontinuities. Furthermore, this attribute can indicate porosity or fluid content in the reservoir [4]. Relative acoustic impedance was calculated by running the integrated the seismic trace through a high-pass Butterworth filter with a hard-coded cut-off at ten-times the sample rate.

The 3D curvature attribute was used in this study to highlight the rate of change of surface shapes mapped as horizons. Tight, short wavelength curvatures delineated details within intense, highly localized fracture systems. Broad, long-wavelength curvatures enhanced subtle flexures on the scale of 100–200 traces [6]. In this study, a vertical radius of seven traces and an inline/crossline radius of one (a comparison between neighbour traces) were chosen for the generation of the minimum curvature.

All seismic attributes were projected on horizon slices for further interpretation and for fault and fracture zone detection.

3.4 Results

3.4.1 Seismic Interpretation

The study area is divided by an elongated Zechstein salt dome striking north–south in the southern part of the seismic survey and turning to east–west orientation in the north. In the Permian interval, $\sim$0.5 to 1.0 km west of the salt dome, six wells were drilled into a structural high bordering the major graben/halfgraben system in the east (Figs. 3.3 and 3.4). The maximum cumulative offset of this NNW to SSE striking, WSW dipping, normal fault is $\sim$900 m, involving the Carboniferous basement. It is composed of a set of subparallel faults associated with relay ramps and pull-apart basins [44]. The fault system is not visible in the area where it is located below the salt dome but its continuation can be mapped in the northern part of the survey. The graben bounding faults on the opposite side, in the west of the interest area, are striking N–S, have a maximum offset of $\sim$700 m and dip towards the east. In the graben area, multiple faults with minor offsets were mapped, the majority running NNW-SSE. Dipping directions vary between ENE and WSW (Fig. 3.4).

The amplitude horizon slices show a pattern of linear features that form a polygonal network, which is often displayed as small linear depression on horizon slices as shown in Fig. 3.5a and b. Except in the north-eastern part (north of the salt wall) the polygonal patterns preferentially occur adjacent to reservoir rocks on Upper Rotliegend II surfaces (Bahnsen/Dambeck, Top Rotliegend/Base Zechstein and Top A2 Zechstein).

By extracting the variance of the surfaces, the resolution of this pattern can be increased. Other attributes such as chaos, intensity, cosine phase and instantaneous frequency show the pattern only partly but not as complete as the variance attribute. Ant tracking in a combination of passive and aggressive tracking with variance as input, displays the pattern on time slices in ~4.500 m depth (Fig. 3.6). It matches the patterns observed on the variance cube slices. Furthermore, the large-scale polygons, which show angles of 80–120°, comprise a pattern of subordinate smaller ones. The minimum and maximum curvature attribute enables an even more detailed distinction of the polygonal network into these smaller units. The polygons of smallest diameter are located just south of the salt dome, increasing in size towards the south (Fig. 3.6). On time slices, the features show a huge contrast in amplitudes compared to the surrounding strata. In the majority of cases, the superordinate polygons are linked to faults with offsets between 35 and 900 m in inlines and crosslines. Frequently, small depressions without offset accompany the fault tops and in some cases no offset can be identified. Irregularities or variances in seismic amplitudes are hardly noticeable at locations of subordinate patterns (Fig. 3.5a and b). On horizon slices, lineaments continue across existing faults. The majority of the lineations appear just south of the salt dome and at least three orders of polygons were interpreted, ranging from diameters of <100 m, close to

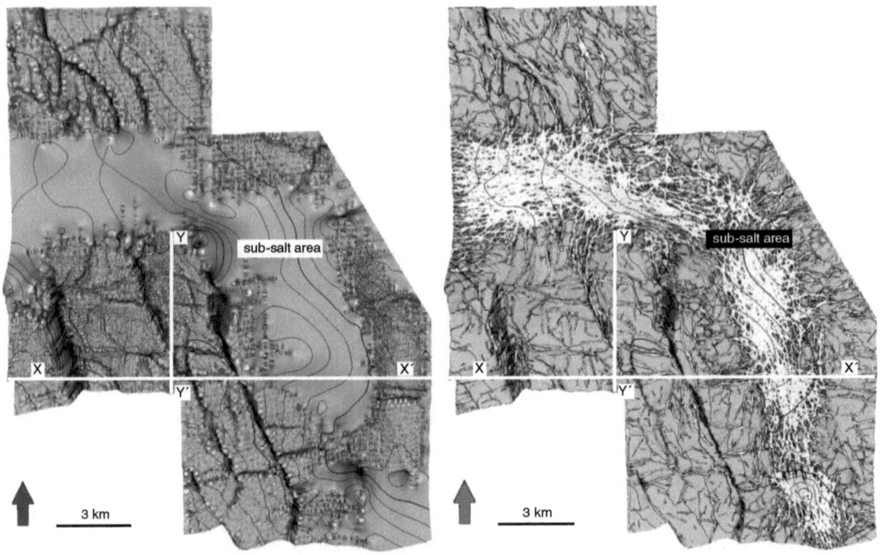

Fig. 3.6 Polygonal pattern on curvature (*left*) and ant tracking (*right*) of Top Rotliegend surface. Acquisition artefacts in the northernmost and southern part on the curvature image. Pattern is displayed in higher detail in Fig 3.7

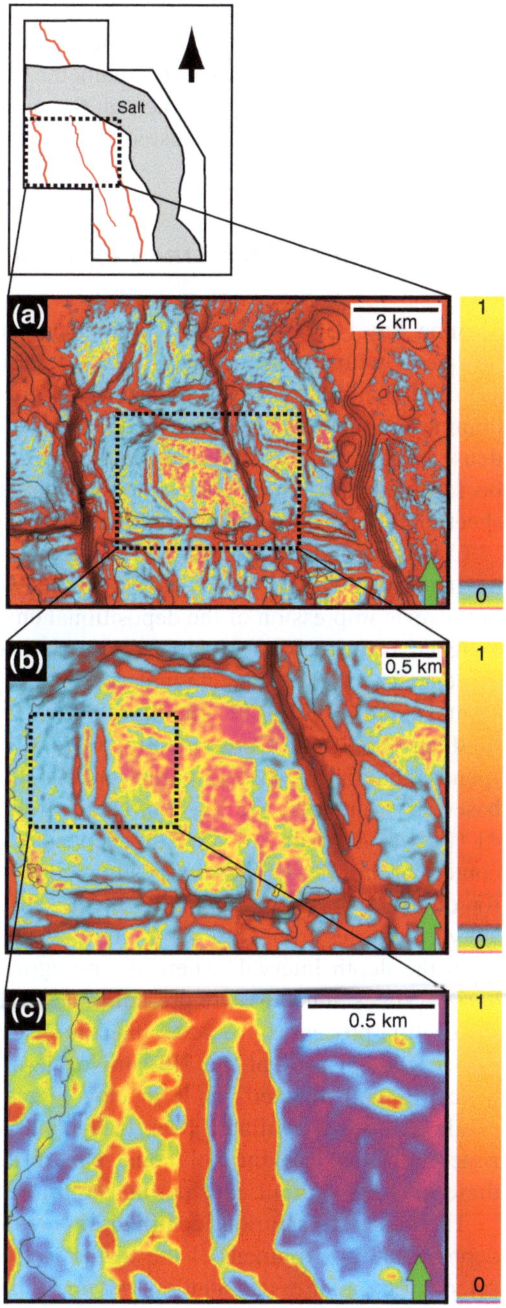

Fig. 3.7 Different scales of the polygonal pattern on the Top Rotliegend surface. **a** major polygons with *red outlines* in the central part of the Top Rotliegend surface. Graben bordering faults with high offsets in the E and W. Subdivision of the Polygen in the center of the image into the *purple* ones with *blue outlines*. **b** Zoom of the central super-ordinate polygon and its subdivision into at least 4 smaller onces with *blue outlines* and *purple fillings*. **c** Zoom of the western outlines of the superordinate polygon. More complex subdivisions with red outlines can be observed just west of the major polygon. Note variation in customised colour templates from *A* and *B* to *C*

the minimum scale that is resolvable with seismic data, to several kilometres, (Fig. 3.7). Further south, the number of WNW-ESE striking lineaments decreases and the NNW-SSE trending features prevail.

Orientation of the lineaments with the highest contrast in variance is NNW to SSE, conformable to the trend of major faults. Angles between the lineations differ from 80 to 120°. The horizontal diameter of the influence of a single lineament ranges 50–250 m. Their influence on the underlying is noticeable for a minimum of 50 ms down from the Zechstein to the reservoir vicinity.

3.4.2 Core and Wire Line Log Data

Gamma Ray (GR) and Sonic curves from the reservoir interval of Well C1 are compared with the sedimentary facies interpretation shown in Fig. 3.8. Gamma Ray and Sonic logs show characteristic maximum peaks corresponding to clay-dominated intervals. Because of the limited vertical resolution of the logs the dense interbedding of different sandflats with varying clay contents can often not be resolved. Minimum GR values are characteristic for intervals consisting of sheetflood deposits and thicker dune successions.

Clay content determination was carried out macroscopically. Due to the varying deviation in Well C2, results do not provide very reliable numbers of clay content, but give an impression of the depositional environment. The total clay content was assigned as the major criteria for the subdivision of sandflat facies. The determined percentages of min/max amounts of clay in the different sedimentary environments: ponds 95–100 %, mudflats 50–95 %, wet sandflat 21–35 % and damp sandflats 10–20 %. Consequently, Core C1 contains 42 % of "wet deposits" (ponds, mudflats, wet and damp sandflats), 44 % "dry deposits" (dunes, dune bases, dry sandflats) and 14 % of channelised fluvial deposits. Core C2 comprises 41 % "wet deposits", 47 % "dry deposits" and 12 % fluvial deposits. The clay content of the wet deposits was estimated at 40–51 % for core C1 and at 35–48 % for core C2. Carbonate contents in the different successions are 0–19.6 % and anhydrite varies from 0 to 4.4 % in Well C2.

In the depth interval, where the polygonal patterns were mapped on seismic horizon slices, mudcurls are present in the cores. These characteristic indicators for desiccation processes provide a hands-on example of mudcracks developed in the Upper Rotliegend II of the SPB in Northern Germany. Furthermore, fluid escape structures and injected sands in clay layers suggest pre-diagenetic dewatering. While no core is available from the polygonal pattern, the large diameter of the zones influenced by the polygonal pattern could cause a problem in identifying these in cores without considering the information provided by seismic data.

The facies distribution identified in cores of Wells C1 and C2 suggest that the depositional environment of the Upper Rotliegend II in the study area was characterized by a large and shallow perennial lake with variable salinities (Fig. 3.8). This lake has been fringed by sand belts that comprise successions from wet, damp and dry sandflats to aeolian dunes with interdune deposits. This setting facilitated the development of kilometre-long cracks, reaching several metres of depth. In the cores, dune successions rarely exceed >2 m thickness. In general, dune

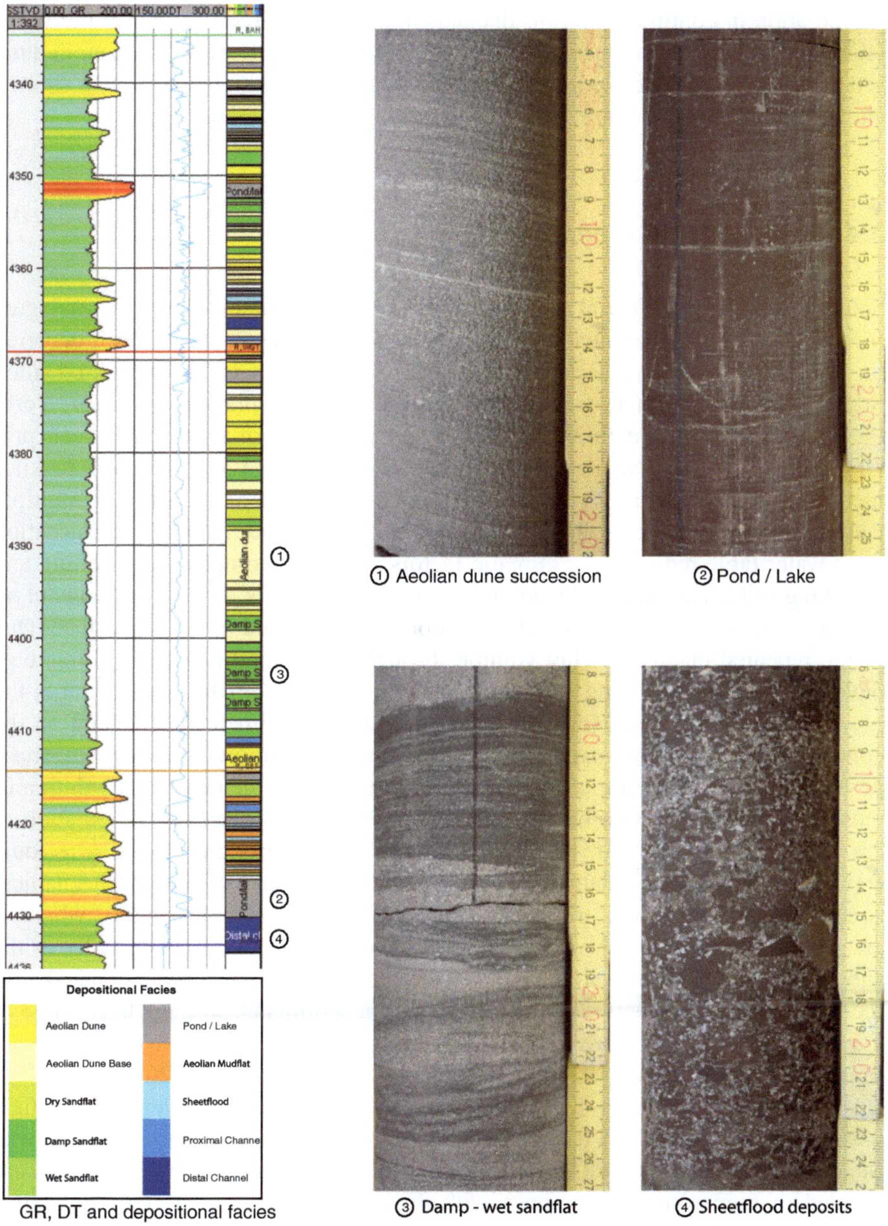

Fig. 3.8 Gamma Ray, Sonic Log and sedimentary facies interpretation of Well C1 plotted on the *left*. Characteristic images of sedimentary facies from the cores on the *right*. Numbers below the images refer to locations in the log

preservation in continental basins depends mainly on the rise and fall of the water table and tectonic subsidence [20, 22]. The presence of numerous barchanoid dune sets indicates that either sediment supply and accommodation space were limited and/or wind velocities were high during the Upper Rotliegend II.

3.4.3 Modern Analogue

Satellite images and aerial photographs provide a suitable modern analogue. Similar patterns of lineaments, often converging to polygons, have been depicted on satellite images and aerial photographs of dry lakes in the western United States. The closest analogues were identified on the playa surface of North Panamint Valley in Death Valley National Park, California and have been previously described by Neal et al. [30] and Messina et al. [26]. Huge, up to kilometre-long cracks with several metres depth developed as a result of multiple wetting and drying cycles close to the sediment–water interface [33]. Desiccation from water table reduction or climatic factors and the accompanying swelling and shrinking of clay minerals played an important role in this genesis. Messina et al. [26] discuss that this type of crack development is initiated in the subsurface. Due to the repeated change in clay volume through drying and dewatering, fractures develop in the surrounding sediment, and the fissures start to grow towards the playa surface. Neal and Motts [29] state that polygon dimensions depend on the ephemeral or long-term nature of desiccation. Furthermore, besides synaeresis processes, earthquakes are often considered to trigger the genesis and evolution of these cracks [26, 33]. On the example of the Panamint Valley, the association of crack endpoints with metre-sized mud volcanoes supports the idea of a tectonic contribution to crack formation. According to Rowell et al [35], abrupt mechanical stress is seen as a major trigger for clay liquefaction and dispersion, which is essential for mud diapir development. The development of cracks initially in the subsurface and finally towards the playa surface, is therefore likely associated with the development of fractures in the surrounding sediments. In their later stage of evolution, the fissures are refilled by aeolian sediment, blown in by wind. Increased growth of vegetation is noticed along the lineaments [26].

Detailed field work in the area reveals that the maximum length of single fissures that build up the polygons in Panamint Valley is approximately 1 km. Superordinate polygons with diameters of some hundreds of metres can be subdivided in numerous smaller ones with diameters of only some tens of metres (Fig. 3.9). Angles of intersections vary from 80 to 120°. Configuration and shape of the polygonal patterns and associated fissures were discussed by Neal et al. [30]. The influence of the wetting and drying effect beneath modern earth fissures was noticed down to a depth of up to 8 m in a study using resistivity measurements in the Chihuahuan Desert [23].

Cracks, which build polygons of different scales, can be observed on satellite images of dry lakes in Arizona, Nevada, California, New Mexico and Oregon,

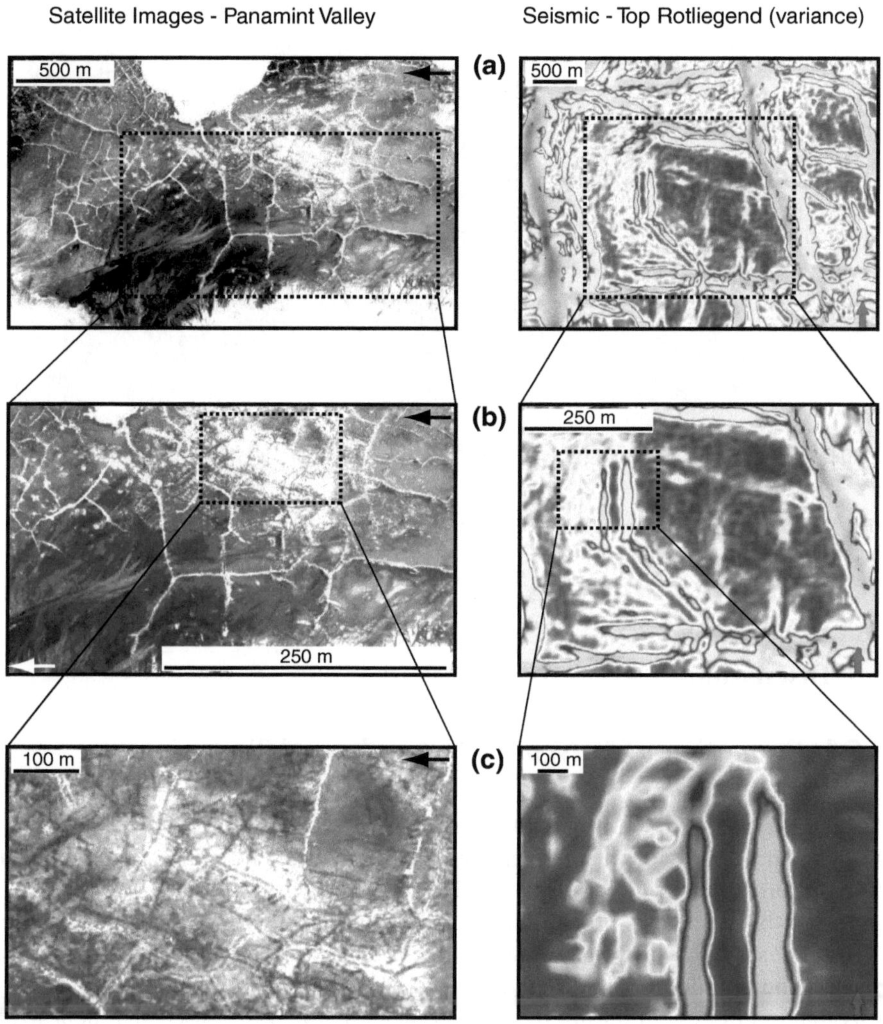

Fig. 3.9 **a** Different scales of polygonal patterns on satellite images of the dry lake in Panamint Valley versus polygonal pattern on the Top Rotliegend surface. **b** Zoom into an area where subdivisions of major polygons are visible. **c** Subdivisions of polygons with diameter of down to tens of metres on satellite images and seismic surface. Image © 2010 Google and DigitalGlobe

United States (Fig. 3.10). Without a certain saline and clay content, the genesis of desiccation cracks in playa lakes is impossible [21]. In this regard, clay content and evaporite content estimated from core analysis (see core and wireline line log data and Fig. 3.8) and during previous studies by Gaupp et al. [11] match the numbers proposed by Neal et al. [30] for the genesis of giant desiccation polygons in Great Basin playas, western United States.

Fig. 3.10 Polygonal pattern on satellite image of Guano Lake, Oregon, U.S. Different dimensions of cracks on the surface of the dry lake. Major polygons are subdivided into smaller ones. Satellite image © 2011 Google, SIO, NOAA, U.S. Navy, NGA, GEBCO

3.5 Discussion

3.5.1 Formation of Polygonal Networks and Tectonics

Although synaeresis is a process that is not fully understood, the model by Cartwright et al. [5] provides a convincing explanation for the formation of a polygonal fault system in three major steps: (1) Heterogeneity in the sedimentary layers is generated by synaeresis processes. (2) The resulting difference in competence of the layers serves as a weakness zone and provides preferred location for stress release and faulting during step two. (3) The weakness zones develop into a polygonal fault system during further sedimentation and increased sediment load.

Applied to this study, Cartwright et al.'s hypothesis implies that, at locations where no clear faults occur or variation in seismic amplitudes can be recognized, and subordinate polygons are only visible on variance, ant tracking and curvature horizon slices, the system is bound to step 1 of the model; alternatively, minor subseismic faults with offsets below vertical seismic resolution may have developed.

Polygonal patterns are also widespread in evaporites, e.g. associated with tepee structures on dry salt lakes. Isopach maps clearly show that the polygonal patterns have not formed during the Zechstein. The polygonal pattern is visible as a negative sediment thickness in the Upper Rotliegend and as a positive thickness anomaly on the A2 Zechstein isopach map [44]. In addition, it is obvious on seismic cross sections that the negative relief of the major polygon outlines developed in the Upper Rotliegend II is evened by the Zechstein sedimentation.

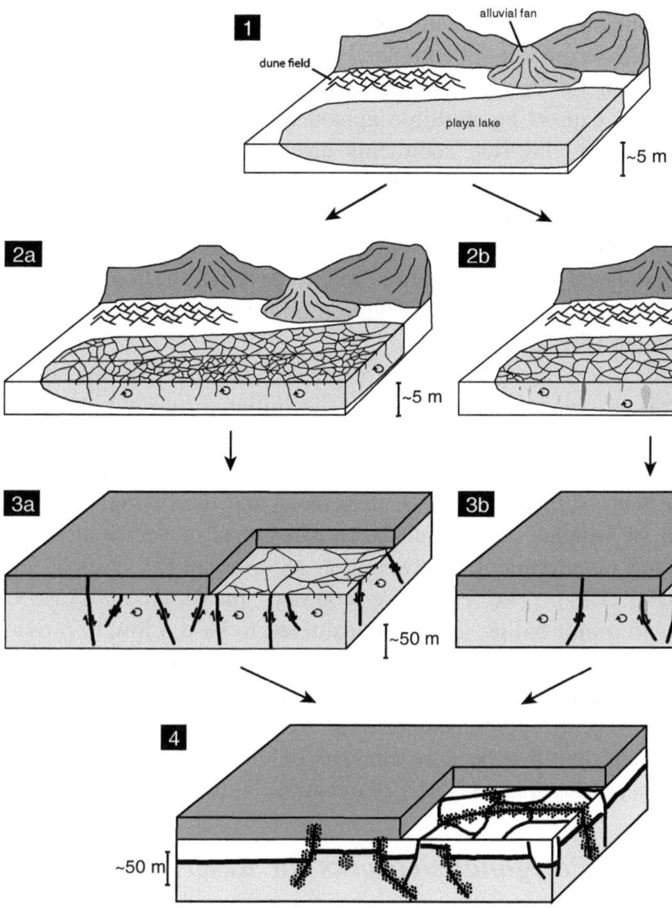

Fig. 3.11 Simplified model for the development of the polygonal pattern depicted on horizon slices in the Upper Rotliegend II of the Southern Permian Basin. (*1*) Initial setting: Dry lake with coastal parallel dune belts. Episodic rainfall but mainly arid climate. (*2a*) During desiccation, cracks develop on lake surface. (*2b*) In contrast fissure generation by synaeresis can start in the subsurface caused by multiple wetting and drying cycles. Cracks develop upwards. Fissures often accompanied by a fractured halo (displayed as *dark* shade). Some reach the surface and cause cracks on the surface. (*3a*) Cracks build polygons on surface; dimensions: cm to km. Cracks are filled by sand, blown in by wind from dunes and clay. Stress increases due to overburden resulting in dewatering of clays. Fluid circulation increases near cracks and in their influenced surrounding, in case of syneresis (*3b*) induced fissures also in the subsurface. (*4*) Competence contrast between crack-filling and surrounding and fractures developed during crack formation serve as preferential locations of fault development. With increasing overburden and stress, faults develop in some locations of former cracks, in certain areas creating a polygonal fault system. In other areas the polygonal fault system remains in the immature state. Serving as fluid pathways an increased cementation took place in some of the faulted/fissured areas (indicated by *point clouds*)

The processes responsible for the development of the kilometre long and metre deep cracks on desert playas are not yet completely understood. Messina et al. [26] state that the evolution of the cracks is initiated by synaeresis and starts in the subsurface with fissures, caused by multiple episodes of wetting and drying of sediments. The swelling of clay-rich sediments has a fracturing effect on the surrounding strata. As a result, a halo-like fractured area develops around first-generation fissures. Early subsurface fissures expand during multiple wetting and drying cycles towards the surface. Some subsurface fissures may break through the playa surface. An increased fluid circulation might develop in fissured areas, which, depending on the processes, causes solution and cementation of minerals during later development (Fig. 3.11).

A second mechanism for the development of cracks on dry lake surfaces is desiccation. In contrast to synaeresis, this process is initiated by mud shrinkage during atmospheric drying of the lake surface ([32]; Fig. 3.11).

After the initial development of cracks at the surface, there is a contrast in competence between the sediment fillings of the cracks and the surrounding lake surface, independent of whether they result from synaeresis or desiccation processes. For further fault development, Cartwright et al.'s model [5] gives a plausible explanation (Fig. 3.11). The evaporite contents measured in core C2 (0–4.4 % anhydrite and minor halite; [9]) are considered to be too low to provide the exclusive mechanism for the development of the polygonal pattern. However, the thermal strain, resulting from volume contraction, might support the processes [21]. A higher initial evaporite content, decreased by later solution during fluid circulation, could not be identified by core analysis or thin sections analysis [9]. A development that includes synaeresis and desiccation is therefore proposed.

3.5.2 Implications of Polygonal Networks for Reservoir Quality

Rotliegend reservoir compartmentalisation through subtle transfer faulting in the Dutch offshore was discussed by Geiss et al. [14]. On the Top Rotliegend horizon slice of the K4–K5 North Sea sectors, Geiss et al. interpreted transfer faults intersecting major faults at angles of 80–120°. These faults show hardly any vertical offset, affecting the Rotliegend, but an offset is visible in the Westphalian. Although these faults are more linear then those in this study area, the degree or lack of vertical displacement is similar.

In the investigated data a negative relief that affects 3–4 seismic reflectors in and crosslines is recognized at locations where fissures can be observed on horizon slices. Hustoft et al. [19] found these associations at fluid migration pathways (sometimes referred to as pipes) in the Miocene Kai Formation, capping polygonal fault systems of the mid-Norwegian margin. The pathways were accompanied by a positive relief $\sim$0.3 ms TWT below the negative bulge. Amplitude variations along the pipes, as they were recognized by Hustoft et al. [19], can not be observed in the data set. Hustoft et al. interpret the pipes to represent pathways for gas with the change in amplitude, resulting from certain gas contents in the tubes. In some cases locations of mud diapirs coincided with such "pull down" effects. In the Panamint Valley, mud diapirs are also located at endpoints of fissure-building macropolygons (see also [26]).

The timing of diagenesis in the northern German Rotliegend II suggests that the general trend of porosity inversion of initially highly porous sediment through enhanced fluid flow and cementation [10, 11] can be applied to the cementation and porosity development of former sandy fissure-fills and weakness zones in this study.

Verschuren [46] reported fluid flow along polygonal faults, based on observations of displaced microfossils in fault gouge. Edwards et al. [6] discussed porosity reduction in the Permo-Triassic Hopeman Sandstone (Moray Firth, Scotland) through enhanced cementation of deformation bands, forming permeability barriers. The degree of cementation and porosity reduction is strongly related to the proximity of a major fault zone. Although deformation bands can not be observed in the cores, the polygonal fissure system might provide fluid pathways, and thus be influenced by fluid migration along a nearby fault zone.

Subtle fissures, hardly resolvable in seismic data, may lead to new strategies for Rotliegend hydrocarbon exploration in the SPB and geologically similar areas worldwide. Assuming that the described polygonal structures develop to subtle migration barriers with increased cementation during a later stage of evolution, they could lead to increased estimations of the recoverable volume in reservoir modelling.

Such a decrease of reservoir quality as described above is recorded in the cores of Well C1 and C2 [10, 11]. In this study the proximity to a fault zone seems to play an important role as reservoir successions of Well C1, located ~300 m closer to the main fault than Well C2 (Fig. 3.4), show lower porosities and permeabilities than stratigraphic equivalent core intervals of Well C2 [10, 11]. In this case, the development of permeability barriers due to increased cementation along original fluid pathways might be the major effect causing the compartmentalisation of the reservoir.

3.6 Conclusions

Results of the sedimentary analysis of cores and logs coincide with the existing depositional model of earlier studies of the Rotliegend in the SPB of northern Germany [18, 31]. In summary, the playa lake environment in Panamint Valley, United States, provides a modern analogue to the Upper Rotliegend II interval in the interest area. Based on this, I conclude that some of the linear depressions that do not show offsets on seismic sections are most likely the result of mud diapirism. The fact that the smallest identifiable polygons are noticed just south of the salt wall matches the results of the minimum palaeorelief reconstruction by Vackiner et al. [44], which identified this region to represent a topographic low and potential depocentre in the interest area. Observations of modern dry lakes on satellite images support the point that the most differentiated networks of cracks are associated with depressions on ephemeral dry lakes.

Based on the presented results, a new model involving lineaments building up polygonal patterns, m to km scale presented on horizon slices of seismic data in the reservoir vicinities and their surroundings at $\sim$4.500 m depth is developed. With the focus on multiple amplitude-dependant attributes to seismic z-slices these patterns are displayed although their lateral extension is below vertical seismic resolution. I propose these pattern to display sub-seismic faults or fissures, whose originally good reservoir quality was inverted during diagenesis through higher cementation. This might have a strong influence on hydrocarbon migration and consequently on hydraulic field connectivity, pressure gradients and hydrocarbon production. Concerning the planning of future well sites, intensive seismic attribute analysis applied to surfaces and z-slices strongly improves the quality of an evaluation process. Furthermore, this new insight into sub-seismic faulting and fracturing helps avoiding difficulties while drilling into highly cemented rocks in future prospects. Regarding reservoir quality prediction, this study provides an approach for re-evaluating tight gas fields in similar areas worldwide.

References

1. Bachmann GH, Hoffmann N (1997) Development of the Rotliegend basin in Northern Germany. Geologisches Jahrbuch D 103:9–31
2. Becker F, Bechstädt T (2006) Sequence stratigraphy of a carbonate-evaporite succession (Zechstein 1, Hessian Basin, Germany). Sedimentology 53:1083–1120
3. Brown AR (2001) Data polarity for the interpreter. Lead Edge 20:549
4. Buxton Latimer R, Davison R, Van Riel P (2000) An interpreter's guide to understanding and working with seismic-derived acoustic impedance data. Lead Edge 19(3):242–256
5. Cartwright J, James D, Bolton A (2003) The genesis of polygonal fault systems: a review. In: Van Rensbergen P, Hillis RR, Maltman AJ, Morley CK (eds) Subsurface sediment mobilization. Geological Society, vol 216. Special Publications, London, pp 223–243
6. Chopra S, Marfurt KJ (2010) Integration of coherence and volumetric curvature images. Lead Edge 29(9):1092–1107
7. Edwards HE, Becker AD, Howell JA (1993) Compartmentalization of an aeolian sandstone by structural heterogeneties: Permo-Triassic Hopeman Sandstone, Moray Firth, Scotland. In: North CP, Prosser DJ (eds) Characterization of fluvial and aeolian reservoirs. Geological Society, vol 73. Special Publications, London, pp 339–365
8. Gast R (1995) Sequenzstratigraphie. In: Plein E (ed) Stratigraphie von Deutschland I; Norddeutsches Rotliegendbecken—Rotliegend-Monographie Teil II. vol 183. Courier Forschungs-Institut Senckenberg, Frankfurt a. M., pp 47–54
9. Gast R, Pasternak M, Piske J, Rasch H-J (1998) Das Rotliegend im nordostdeutschen Raum: Regionale Übersicht, Stratigraphie, Fazies und Diagenese. Geologisches Jahrbuch A 149:59–79
10. Gaupp R, Fischer C (2000) Diagenese der Oberrotliegendsandsteine aus Bohrungen. Not published
11. Gaupp R, Solms M (2005) Sedimentological and petrological investigations. In: Paleo oil- and gasfields in the Rotliegend of the North German basin: effects upon hydrocarbon reservoir quality (Paläo-Öl- und Gasfelder im Rotliegenden des Norddeutschen Beckens: Wirkungen der KW-Migration auf die Speicherqualitäts-Entwicklung). DGMK Research Report, 593–8: Tight gas reservoirs—natural gas for the future. pp 1–44
12. Gaupp R, Matter A, Platt J, Ramseyer K, Walzebuck JP (1993) Diagenesis and fluid evolution in deeply buried Permian (Rotliegende) gas reservoirs, Northwest Germany. AAPG Bull 77(7):1111–1128

13. Gebhardt U, Plein E (1995. Neue Gliederung. In: Plein E (ed) Norddeutsches Rotliegendbecken, Rotliegend-Monographie Teil II, vol 183. Courier Forschungs-Institut Senckenberg, Frankfurt a. M., pp 18–24
14. Gebhardt U, Schneider J, Hoffmann N (1991) Modelle zur Stratigraphie und Beckenentwicklung im Rotliegenden der Norddeutschen Senke. Geologisches Jahrbuch A 127:405–427
15. Geiss B, Kremer Y, van Koppen J, Bertotti G (2009) Field compartmentalisation through subtle transfer faulting: an example from blocks K4/K5 offshore NL. 71st EAGE Conference and Exhibition, Amsterdam, 8–11 June
16. George GT, Berry JK (1993) A new palaeogeographic and depositional model for the Upper Rotliegend of the UK Sector of the Southern North Sea. In: North CP, Prosser DJ (eds) Characterization of fluvial and aeolian reservoirs. Geological Society, vol 73. Special Publications, London, pp 291–319
17. Glennie KW (1972) Permian Rotliegendes of Northwest Europe interpreted in light of modern desert sedimentation studies. AAPG Bulletin 56(6):1048–1071
18. Glennie KW (1983) Early Permian (Rotliegendes) palaeowinds of the North Sea. Sed Geol 34:245–265
19. Glennie KW (1986) Development of N.W. Europe's Southern Permian gas basin. In: Brooks J, Goff JC, Van Horn B (eds) Habitat of Paleozoic gas in N.W. Europe. Geological Society, vol 23. Special Publications, London, pp 3–22
20. Hustoft S, Mintert J, Bünz S, Nouzé H (2007) High-resolution 3D-seismic data indicate focussed fluid migration pathways above polygonal fault systems of the mid-Norwegian margin. Mar Geol 245:89–106
21. Kocurek G (1999) The aeolian rock record (Yes, Virginia, it exists, but it really is rather special to create one). In: Goudie A, Livingstone I (eds) Aeolian environments sediments and landforms. Wiley, Chichester, pp 239–259
22. Kocurek G, Hunter RE (1986) Origin of polygonal fractures in sand, uppermost navajo and page sandstones, page Arizona. J Sediment Petrol 56(6):895–904
23. Kocurek G, Robinson NI, Sharp JM (2001) The response of the water table in coastal aeolian systems to changes in sea level. Sed Geol 139:1–13
24. Langford RP, Eshete T (2000) Fissure behaviour in the Chihuahuan desert, and depth estimation using sediment strength. Environ Eng Geosci VI(4):301–309
25. Legler B (2005) Faziesentwicklung im Südlichen Permbecken in Abhängigkeit von Tektonik, eustatischen Meeresspiegelschwankungen des Proto-Atlantiks und Klimavariabilität (Oberrotliegend, Nordwesteuropa). Schriftenreihe der Deutschen Gesellschaft für Geowissenschaften, 47
26. Lohr T, Krawczyk CM, Tanner DC, Samiee R, Endres H, Thierer PO, Oncken O, Trappe H, Bachmann R, Kukla PA (2008) Prediction of subseismic faults and fractures. Integration of three-dimensional seismic data, three-dimensional retrodeformation, and well data on an example of deformation around an inverted fault. AAPG Bull 92(4):473–485
27. Messina P, Stoffer P, Sith WC (2005) Macropolygon morphology, development, and classification on North Panamint and Eureka playas, death valley national park CA. Earth Sci Rev 73:309–322
28. Mohr M, Kukla PA, Urai JL, Bresser G (2005) Multiphase salt tectonic evolution in NW Germany; seismic interpretation and retro-deformation. Int J Earth Sci 94(5–6):917–940
29. Mohr M, Warren JK, Kukla PA, Urai JL, Irmen A (2007) Subsurface seismic record of salt glaciers in an extensional intracontinental setting (late Triassic of northwestern Germany). Geology 35:96–966
30. Neal JT, Motts WS (1967) Recent geomorphic changes in playas of the western United States. J Geol 75:511–525
31. Neal JT, Langer AM, Kerr PF (1968) Giant desiccation polygons of great basin playas. Geol Soc Am Bull 79(1):69–90

32. Plein E (1995) Stratigraphie von Deutschland I: Norddeutsches Rotliegendbecken, Rotliegend-Monographie Teil II, Courier Forschungs-Institut Senckenberg, Frankfurt a. M., 183
33. Plummer PS, Gostin VA (1981) Shrinkage cracks: desiccation or synaeresis? J Sediment Petrol 51:1147–1156
34. Pratt BR (1998) Synaeresis cracks: subaqueous shrinkage in argillaceous sediments caused by earthquake-induced dewatering. Sed Geol 117:1–10
35. Rieke H (2001) Sedimentologie, Faziesarchitektur und Faziesentwicklung des kontinentalen Rotliegenden im Nordostdeutschen Becken (NEDB). Ph.D. thesis, Potsdam University, Germany
36. Rowell DL, Payne D, Ahmad N (1969) The effect of the concentration and movement of solutions on the swelling, dispersion, and movement of clay in saline and alkali soils. J Soil Sci 20(1):176–188
37. Schneider J, Gebhardt U, Gaitzsch B (1995a) Fossilführung und Stratigraphie. In: Plein, E. (ed.), Norddeutsches Rotliegendbecken, Rotliegend-Monographie Teil II. vol 183. Courier Forschungs-Institut Senckenberg, Frankfurt a. M. pp 25–39
38. Schwarzer D, Littke R (2007) Petroleum generation and migration in the, 'tight gas' area of the Germany Rotliegend natural gas play: a basin modelling study. Pet Geosci 13:37–62
39. Stemmerik L (2001) Sequence stratigraphy of a low productivity carbonate platform succession: the upper Permian Wegener Halvø formation, Karstryggen area, East Greenland. Sedimentology 48:79–97
40. Stollhofen H (2007) Postvulkanische Rotliegend-Schwemmfächersysteme am Hunsrück-Südrand, Saar-Nahe-Becken, SW-Deutschland. Jahresberichte und Mitteilungen des Oberrheinischen Geologischen Vereins, 89:285–306
41. Stollhofen H, Bachmann GH, Barnasch J, Bayer U, Beutler G, Franz M, Kästner M, Legler B, Mutterlose J, Radies D (2008) Upper Rotliegend to early cretaceous basin development. In: Littke R, Bayer U, Gajewski D, Nelskamp S (eds) Dynamics of complex intracontinental basins. The central European basin system. Springer, Berlin, pp 181–210
42. Strohmenger C, Voigt E, Zimdars J (1996) Sequence stratigraphy and cyclic development of basal Zechstein carbonate-evaporite deposits with emphasis on Zechstein 2 off-platform carbonates (Upper Permian, Northeast Germany). Sed Geol 102:33–54
43. Strömbäck AC, Howell JA (2002) Predicting distribution of remobilized aeolian facies using sub-surface data: the Weissliegend of the UK Southern North Sea. Pet Geosci 8:237–249
44. Vackiner A, Antrett P, Back S, Kukla P (2010) Sedimentological and structural analysis of a Rotliegend tight gas field in NW Germany. GMK/ÖGEW-Frühjahrstagung 2010, Fachbereich Aufsuchung und Gewinnung, Celle, 12–13 April
45. Vackiner A, Antrett P, Stollhofen H, Back S, Kukla PA, Bärle C (2011) Syndepositional tectonic controls and Palaeo-topography of a Pemrian tight gas reservoir in NW Germany. J Pet Geol 34(4):411–428
46. Van Wees J-D, Stephenson RA, Ziegler PA, Bayer U, McCann T, Dadlez R, Gaupp R, Narkiewicz M, Bitzer F, Scheck M (2000) On the origin of the Southern Permian basin, central Europe. Mar Pet Geol 17:43–59
47. Verschuren M (1992) An integrated 3D approach to clay tectonic deformation. Unpublished Ph.D. Thesis, Universiteit, Gent, Belgium
48. Voigt T, Reicherter K, von Eynatten H, Littke R, Voigt S, Kley J (2008) Sedimentation during basin inversion. In: Littke R, Bayer U, Gajewski D, Nelskamp S (eds) Dynamics of complex intracontinental basins. The central European basin system. Springer, Berlin, pp 211–232
49. Warren JK (2008) Salt as sediment in the central European basin system as seen from a deep time perspective. In: Dynamics of complex intracontinental basins; the central European basin system. Springer, Berlin, Federal Republic of Germany, pp 249–276
50. Ziegler PA (1982) Geological atlas of Western and central Europe. Shell Internat. Petrol Maatsch, The Hague

Chapter 4
Impact of Arid Surface Mega-Cracks on Hydrocarbon Reservoir Properties

4.1 Introduction

The identification of compartmentalised reservoirs and the understanding of potential controls on their formation plays an important role in developing hydrocarbon fields. However, it is not always possible to predict reservoir compartmentalisation based solely on seismic data. Although information from cores and wire line logs improve our understanding of compartmentalisation, rapid decline of production rates can not often be satisfactorily explained or predicted. Common solutions for maintaining production rates involve the drilling of additional, expensive wells and the application of hydraulic fracturing techniques. A better understanding of how a reservoir is compartmentalised could allow for more efficient well and completion designs.

In this study, reasons for the compartmentalisation of a Rotliegend II (Upper Permian) tight gas field in northern Germany (Fig. 4.1) are investigated. The reservoir rocks in this field are of fluvio-aeolian origin and were deposited in the marginal areas of a playa lake under arid conditions [38]. Seismic analysis of the Rotliegend reservoir reveals the presence of conspicuous polygonal pattern which potentially channelise fluid flow and thus may control reservoir connectivity.

The multi-directional pattern addressed as polygonal pattern in this study does not represent a typical polygonal pattern as it is known from attribute maps of polygonal fault systems from the Tertiary of the North Sea [8] or the opal-A to opal-CT boundary from offshore Norway [5] in shallower depths. The term furthermore describes the pattern that is visible on TWT surfaces composed of the outlines of faults and sub-seismic features of varying orientations.

To gain a better understanding of the origin and evolution of the Rotliegend polygonal fracture systems, the Panamint Valley in California, near Death Valley has been chosen as a modern field analogue where modern, large-crack systems are directly accessible at the dry lake surfaces. Similar to their ancient analogues, the modern polygonal crack systems extend laterally over several kilometres (miles) with a maximum polygon diameter of several metres and a maximum depth of >1 m (3 ft), making them clearly visible on satellite images of the Panamint lake surface.

P. Antrett, *Characterization of an Upper Permian Tight Gas Reservoir*,
Springer Theses, DOI: 10.1007/978-3-642-36294-1_4,

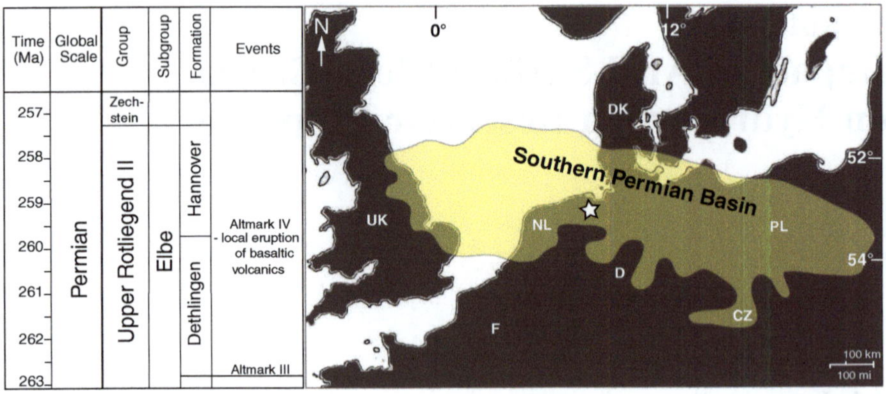

Fig. 4.1 Stratigaphic chart of the Upper Permian in Northern Germany on the *left* (modified from [30], timescale according to [11]. *Right* maximum extent of the Southern permian basin (modified after [43]. Study area displayed in Fig. 4.2 marked by the *star*. *D* Germany, *F* France, *NL* Netherlands, *PL* Poland, *UK* United Kingdom

4.2 Depositional Settings

4.2.1 Tight Gas Field Study Area

The Upper Rotliegend II strata in northern Germany were deposited in a compartment of the E–W trending SPB extending from the United Kingdom across the southern North Sea through northern Germany into Poland and the Baltic States. Its width varies between 300 and 600 km ($\sim$190 and 370 miles; Fig. 4.1). The basin centre was located in Northern Germany. As the basin developed, the depositional area enlarged from the centre towards the east and west through time [35]. At its maximum extent during the Late Upper Rotliegend II, the basin covered an area of $\sim$429,000 km^2 ($\sim$170,000 miles^2).

Sedimentation during the Upper Rotliegend II in the SPB was dominated by siliciclastics and minor evaporites. The arid to semi-arid climate controlled the depositional environment [16]. Ephemeral fluvial (wadi), aeolian, sabkha and lacustrine environments were identified as the four major facies associations by George and Berry [15], Strömbäck and Howell [36], Legler [24]. Glennie [17], Gast et al. [12], Rieke [32] stated that a huge amount of sediment was supplied through aeolian transport by strong winds from the east. Furthermore, ephemeral fluvial systems draining into the SPB were sourced by the Variscan Orogen and local highs towards the south.

Initiated during deposition of the Dethlingen Formation (Fig. 4.1), a perennial saline lake located in the centre of the SPB had a major impact on base level changes. As a result of its low relief, even small lake level variations caused considerable lateral shifts in depositional facies. The record of these shifts is expressed in two third-order tectonic cycles, divided by the Altmark III and IV

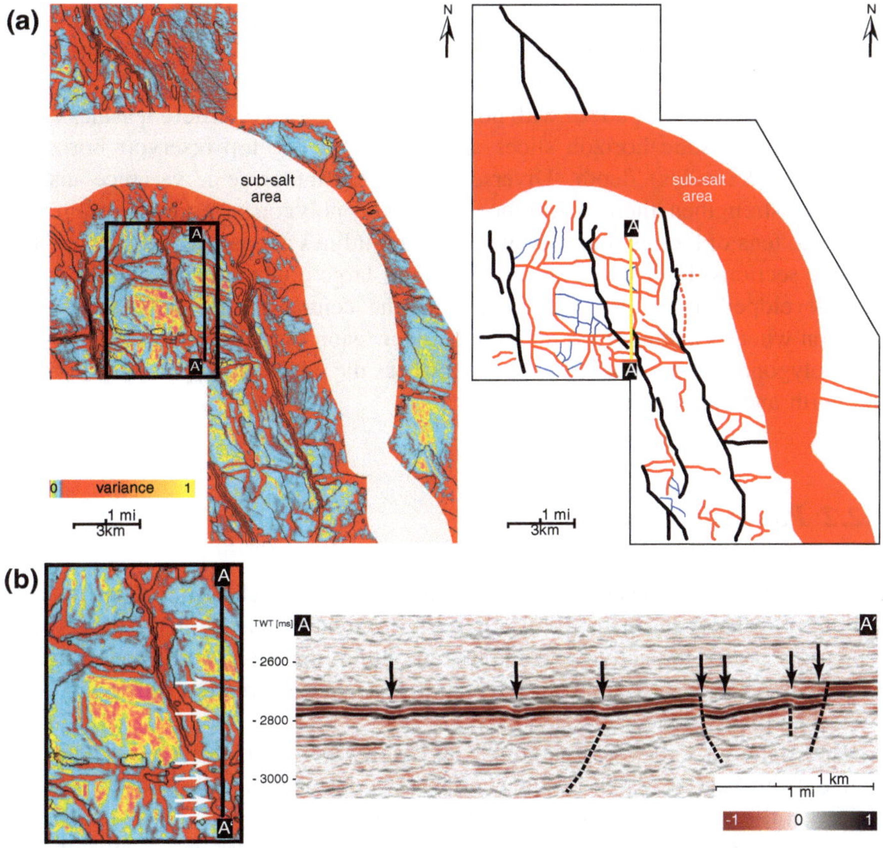

Fig. 4.2 **a** Overview of the German study area. Variance map displayed on the *left*. Schematic sketch on the *right* major faults in *black*. Minor faults in *red* and additional linear features that can not be identified in seismic cross sections in *blue*. **b** Zoom on profile A–A′ location on the *left*. Cross section A–A′ on the *left*. *Arrows* mark locations of the polygonal pattern. A significant offset is not visible in all locations

tectonic events, consisting of 14 fourth-order transgressive and regressive cycles, each with a duration of ~400 ka [11] during the deposition of the Elbe Subgroup (Fig. 4.1). Lake level highstands can be identified by the deposition of claystones. Lake level lowstands and insolation are recorded by halite sedimentation. The ephemeral lake margins were fringed by sand belts comprising successions of wet, damp and dry sandflats and aeolian dunes with interdune deposits.

The tight gas field is located at the south-western margin of the SPB. During the Elbe Subgroup the sedimentation in the area was dominated by dunes, sandflats and saline mudflats but without a permanent connection to the great perennial lake at the centre of the SPB [18, 42]. Vackiner [38] showed evidence of Upper Rotliegend fault activity in the area of interest based on isopach maps. These maps clearly indicate maximum sediment thickness contrasts of ~250 m (820 ft) between hanging- and footwall of N–S and NNW–SSE trending fault zones. Furthermore, a basin that served as a depo-centre during the Upper Rotliegend was identified in the central study area [38].

A multi-directional pattern with polygonal symmetries of different scales can be observed on seismic horizon slices extracted along the top reservoir horizon at $\sim$4.5 km (2.8 miles) depth. Diverse seismic attributes, e.g. variance and ant tracking, strengthen their occurrence. Observed polygon sizes range from kilometres to tens of metres. Large-scale polygon-outlines are associated with faults in seismic sections (Fig. 4.2). The subdivision of large polygons into smaller ones is best developed and most sophisticated in the centre of the study area, at the location where Vackiner [38] identified a depression during the upper Rotliegend. The polygonal pattern does not prevail towards the edges of the seismic survey, to the north and south (see Chap. 3).

4.2.2 Panamint Playa

Panamint Playa is located in the Panamint Valley, California, a north–south striking pull-apart basin in the south-western Great Basin, United States, which represents one of the active grabens that lie within the Basin and Range Province [34]. It is located in a zone of active transtensive deformation along the North American plate boundary [31].

The Panamint Valley is separated from Death Valley in the east by the Panamint Range, from Searles Valley to the southwest by the Slate Range and from Owens Valley in the west by the Argus and Coso Ranges ([34]; Fig. 4.3). Holocene slip rates for the Panamint Valley Fault Zone, based on offset drainages developed on Holocene alluvial fans by Zhang et al. [40] indicate a slip rate of 2.5 $\pm$ 1 mm/year (0.098 $\pm$ 0.04 in/year). Similar values of 2.5–2.7 mm/year (0.098–0.107 in/year) long-term slip rate at the Hunter Mountain Fault Zone (Fig. 4.3) from the Pliocene are reported by Burchfiel et al. [3]. Densmore and Anderson [7] postulate a late Quaternary slip rate of 0.4–0.5 mm/year (0.016–0.02 in/year) for the Ash Hill Fault Zone (Fig. 4.3). Since 1982, the Unite States Geoligical Survey Northern California Seismic Network (USGS NCSN) Catalog (http://www.ncedc.org/) recorded $\sim$3,286 earthquakes in an area 50 $\times$ 50 km around Panamint Springs. The maximum magnitude was 5.2. However, most of the quakes (3005/3286) ranged between magnitudes 1 and 2. Only $\sim$12 earthquakes with magnitudes of 2.3–3.2 were located in the northern part of the Panamint Valley. The hard and smooth surface of Panamint Playa dry lake at an elevation of $\sim$460 m (1,509 ft) above mean sea level is composed of clay, silt, and minor fine-grained sand [27, 34]. It comprises more plastic lacustrine sediments of Pleistocene Lake Panamint at depth. Panamint Playa represents a typical dry lake of the south-western United States, with typical geomorphic features such as solution depressions, drain holes, mud volcanoes,

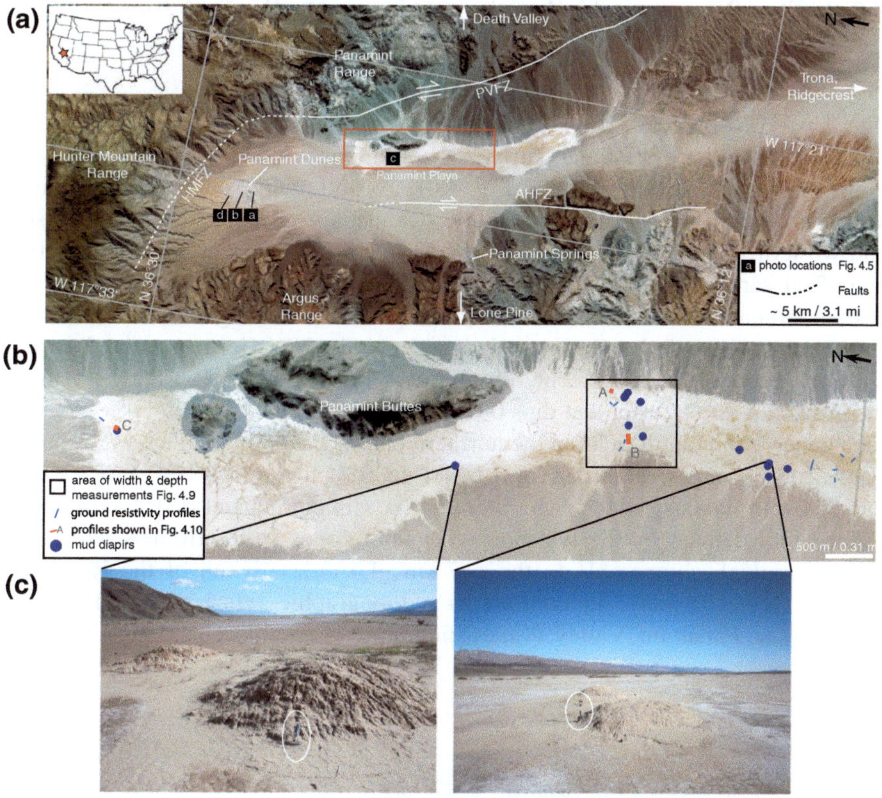

Fig. 4.3 **a** Overview of the Panamint valley including mountain ranges and major fault zones: *PVFZ* Panamint valley fault zone, *HMFZ* Hunter mountain fault zone, *AHFZ* Ash hill fault zone (after [22]). Location of *b* framed in *red*. **b** Zoom of northern part of Panamint playa. Ground resistivity profiles, mud diapirs and photo locations of Fig. 4.10 highlighted (see key for description). **c** Two pairs of mud diapirs on Panamint playa. Satellite images © 2010 Google and Digitalglobe

phreatophyte and spring mounds [26]. However, the most striking features are laterally extensive (>1 km, >0.62 miles) and deep cracks (>1 m, >3 ft) on the playa. Neal and Motts [28] discussed variations in the groundwater table, piezometric surfaces, capillary effects and soil moisture as playing important roles in crack development. According to Messina et al. [26], synaeresis effects might initiate the development of these mega-crack systems. The diameter of the polygons between the cracks ranges from metres to hundreds of metres (several feet to thousands of feet). A 4 m (~ 12 ft) drill core, just south of the Panamint Buttes, shows predominant clay intervals in the lower 2.75 m (9 ft) and slightly coarser grained intervals of fine sandy silt and silty fine sand in the upper 1.25 m (4.1 ft; [21]).

4.2.3 The Panamint Valley as an Upper Rotliegend Field Analogue

The Upper Rotliegend II basin in NW Germany is bordered by NNW–SSE trending faults in the west and partly in the east, constituting a ~7 km (4.5 miles) wide graben/halfgraben system (Fig. 4.4). A pull-apart sub-basin with higher sediment thicknesses between antithetic oversteps is associated with the eastern and western fault according to Vackiner [38]. In addition, variable dip directions and offset variations mapped along the fault in the centre of the graben/halfgraben system indicate a left-lateral strike slip component ([38]; Fig. 4.4). Similar to the structural setting of the Rotliegend depocenter in NW-Germany, the Panamint Valley is bordered by the steeply dipping Ash Hill Fault Zone in the west and by the Panamint Fault Zone, that dips 0–15° W [3, 19] to the east (Fig. 4.3). Close to the Hunter Mountain Range, in the northern part of Panamint Valley, the Panamint Fault Zone changes its strike from NNW to SSE towards WNW–ESE, accompanied by an increase of the right lateral strike slip component ([3]; Fig. 4.3). The basin width defined by the two bordering fault zones varies from ~7 to 12 km (4.5–7.5 miles), comparable to the ~7 km (4.5 miles) width of the Rotliegend basin in NW-Germany. In contrast to a topographic relief of up to 1,700 m (5,580 ft) at Panamint Valley (valley floor to top of adjacent mountain range),

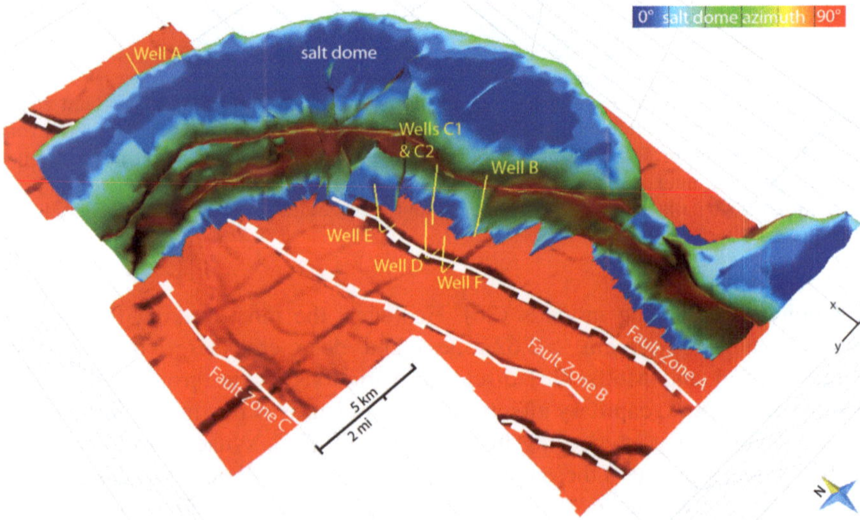

Fig. 4.4 Graben/halfgraben system of the German subsurface study area. Top Rotliegend surface (TWT) displayed in *red*. Salt dome, wells and fault zones *A* to *C* sketched. fault zones were partly active during the Upper Rotliegend. According to Vackiner [38], the offset of fault zone *A* adjacent to the well locations C1 to F developed during the Triassic to Cretaceous and did not exist during the Upper Permian

the palaeorelief associated with the Rotliegend basin in Germany were less pronounced, approaching a maximum of ~250 m (820 ft; Vackiner [38]).

A playa lake, dominated by clay deposition, is situated at the centre of northern Panamint Valley. Striking features are huge cracks on the playa surface reaching lengths of >1 km and depths of >1 m (3 ft). Mud diapirs, up to 1.2 m (3.9 ft) high and up to 2 m (6 ft) in diameter, occur in pairs and are distributed randomly over the playa surface (Fig. 4.3). Alluvial fans descending from the Panamint Range in the east and the Argus Range in the west provide coarse-grained material that is shed over the playa lake surface at its borders. The sedimentary detritus shed from the east and west comprises a high amount of basaltic to andesitic volcanics.

In cores from the Upper Rotliegend II in Northern Germany, stacked dune deposits have maximum thicknesses of ~2.5 m (8.2 ft) with varying dip directions, interpreted to indicate barchanoid dunes and amalgamated dunes [38], very similar to the star to barchanoid dunes in the Panamint Valley. Based on the facies distribution in cores, we concluded [1] that the depositional environment of the Upper Rotliegend II in the area of interest was characterized by the existence of a large but shallow perennial lake, fringed by sand belts, comprising successions from wet, damp and dry sandflats to aeolian dunes with interdune deposits (Fig. 4.5). Possible mud diapirs were identified on 3D seismic data in the northern German Upper Rotliegend II, which are largely similar in appearance to mud diapirs associated with a polygonal fault systems of the Miocene Kai Formation offshore Norway [20].

4.3 Data and Methods

At Panamint Valley, modern crack shapes and geometries were studied by measuring ground resistivity profiles across the cracks. The advantage of the geoelectrical measurements compared to surface observations is the ability to display the whole area influenced by crack evolution due to the differences in resistivity. In this study, varying water contents of different sediments (clay to sand) result in the resistivity variations mapped in the profiles. The resolvable interval ranges from the surface down to the groundwater table, below which the resistivity contrast is diminished due to high water saturation [25]. In addition to the vertical extent, the horizontal extent influenced by the crack can be determined from resistivity measurements, because cracks contain more water than the surrounding sediments. For the measurements, the “Lippmann 4-point light” geoelectrical instrument was used applying the dipole–dipole configuration with a maximum of 80 electrodes [41]. Electrode spacing varied between 1.0 m (3 ft) and 0.25 m (0.8 ft) to achieve (a) either a high resolution or (b) to collect information from greater depth. Data was processed using the Res2DInv software (http://www.geoelectrical.com). Satellite images and surveyed crack width and depth data were considered to determine preferred crack orientations and/or possible links to fault activity. Sediment samples were taken from the cracks and their vicinity for x-ray

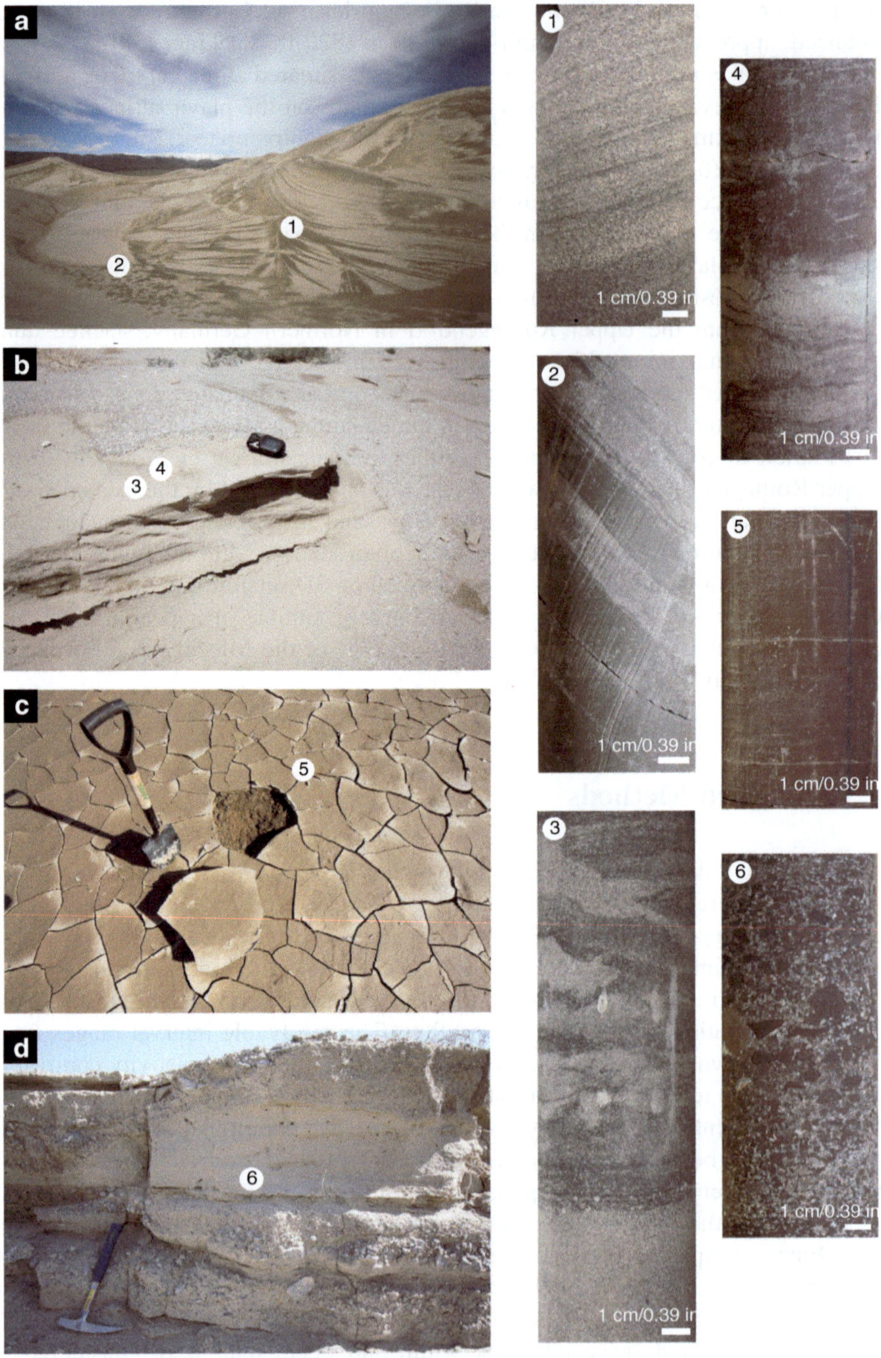

Fig. 4.5 Comparison of sedimentary facies of the Panamint valley and the study area in Germany. Photo locations a–d marked in Fig. 4.3. Core photos of the Wells C1 and C2: (*1*) dune deposits, (*2*) pond margin deposits with dune deposits migrating into the pond. Note the deviation. (*3*) damp sandflat, (*4*) wet sandflat, (*5*) mudflat, (*6*) low energy fluvial deposits

diffractometry analysis (XRD). Swelling clays were identified by glycol dehydration measurements. Results were compared to sediment composition of equivalent sediment facies from cores of the Rotliegend tight gas reservoir in northern Germany.

For the Rotliegend tight gas field in Germany, a 3D seismic survey covering 293 km^2 (114 $miles^2$) with a trace length of 5 s and a sample interval of 4 ms is available. The bin size is 25 × 25 m (82 × 82 ft). Seismic data are displayed in European polarity: an increase in impedance gives negative amplitudes, displayed as trough [2]. For further investigations, the variance seismic attribute was used. With the variance attribute, discontinuities in the horizontal continuity of amplitudes can be displayed as they occur along faults or depositional features. With the variance attribute discontinuities in the horizontal continuity of amplitudes can be displayed as they occur along faults or depositional features. Additionally, a depth-converted version of the seismic survey is available. Furthermore, wireline logs from 6 wells, core material from four wells and Formation Micro Scanning/ Formation Micro Imaging (FMS/FMI) data from one well are taken into consideration.

4.4 Polygonal Mega-Crack Systems

A polygonal crack pattern is obvious on satellite images of Panamint Valley (Figs. 4.3b, 4.6, 4.7). The pattern comprises lineaments, interfering at angles between 80 and 120°. Polygon diameters range from several hundreds of metres to metres. In some parts of the playa, major polygons can be subdivided into smaller ones. At least three dimensions of polygons, with diameters ranging from kilometres to metres, can be identified on satellite images. From the Panamint Buttes (Fig. 4.3), the lineaments are clearly identifiable as cracks or bulges, often accompanied by lines of vegetation (Fig. 4.7).

Satellite images from 1994, 2005 and 2009 indicate that the positions of the mega-cracks on the playa surface are stable at least during this period of 15 years (Fig. 4.6). New crack development seems to be the exception.

Detailed investigation of the cracks on the playa surface revealed at least four different varieties of cracks (Fig. 4.7). The first type comprises subsurface collapse structures (Fig. 4.7a). These subsurface structures became exposed as the surface collapsed due to additional weight. The second type is characterized by vegetation growing in the cracks (Fig. 4.7d). During our stay, the cracks were filled with water for more than 4 weeks after the last rain. Due to evaporation salt rims develop. A wet area, including the salt rims, accompanies the cracks and I refer to this as the "influence area" (Fig. 4.7b–e). With time, sand sourced by the nearby Panamint Dunes gets blown into the cracks and subsequently refills them. Vegetation growing at the edges or in the cracks (Fig. 4.7d) serve as wind barriers and sediment traps and supports this process. After a heavy sandstorm a sandfill of >10 cm (4 in.) in a fresh crack was observed. The third crack type is represented

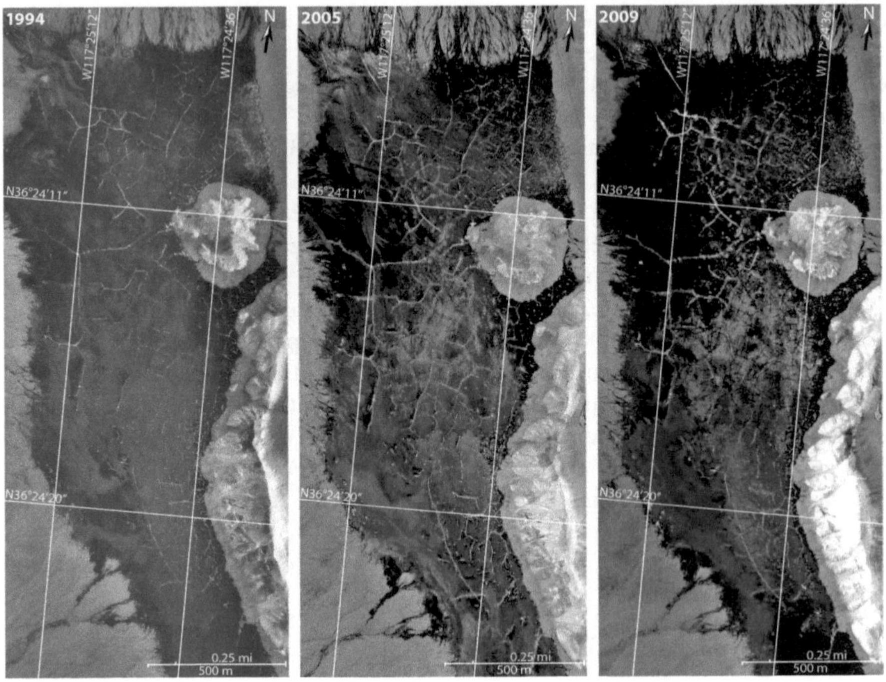

Fig. 4.6 *Black* and *white* colour inverted satellite images of the northern part of the Panamint Playa from 1994 to 2009. Locations of mega-cracks stable over 15 years (images modified from © 1994 U.S. Geological Survey, © 2005, 2009 Google)

by refilled cracks. In some cases bushes still grow on the former crack most probably because the porous sandfill contains available water in contrast to the surrounding clay sediments (Fig. 4.7e). This is confirmed by the resistivity measurements. The accumulated sand forms bulges, which may grow to a height of up to 0.5 m (1.5 ft; Fig. 4.7f). Other examples of filled cracks without a relief developed were only identifiable by higher moisture content and the coarser grain size compared to the surrounding on the playa surface (Fig. 4.7g). In 14 cases, mud diapirs are associated with cracks or exist in the vicinity of the cracks on the playa surface (see Fig. 4.3). The fourth variety of cracks was described as ring fissures by Neal et al. ([29]; Fig. 4.7h). These circular features often show a depression in the centre, where water gathers after rainfall. Some of the ring fissures are connected to collapsed cracks, to cracks containing vegetation and to refilled cracks. In some cases triple junctions developed. However, in most cases, there is no connection to cracks visible at the surface.

Some of the crack orientations match the trends of pre-quaternary faults (Fig. 4.8). To examine a potential relationship between faulting and crack formation, crack orientations, crack width and depth were systematically surveyed (Fig. 4.9). No preferred crack orientation is obvious, nor do particular crack orientations develop enhanced width. In general, the ratio of width versus depth of

Fig. 4.7 Overview of the appearance variety of surface cracks: **a** subsurface development. **b** Recently developed crack. **c** Vegetation growth on crack rim. **d** Vegetation growth in cracks, influence area as halo along crack. **e** Refilled crack. Enhanced vegetation growth on crack-fill. Influence area (wet) along refilled crack. **f** Sandy bulge on refilled crack. Bushes provide sediment trap. **g** Higher moisture content of refilled crack in contrast to remaining playa surface. **h** Ring fissure. Vegetation growth on the rim; depression in the centre

open cracks (cavities) varies between 2 and 4 with a mean value of 3. The maximum measured crack depth is 1 m (3.281 ft); maximum width recognized at a triple-junction is 2.5 m (8.2 ft). There is a linear relationship between crack width and depth (Fig. 4.9). Values of the influence area rarely exceed 7 m (23 ft).

Apart from the contrasting grain size distributions between crack-fill and remaining playa lake surface, XRD analysis reveals no striking difference in the

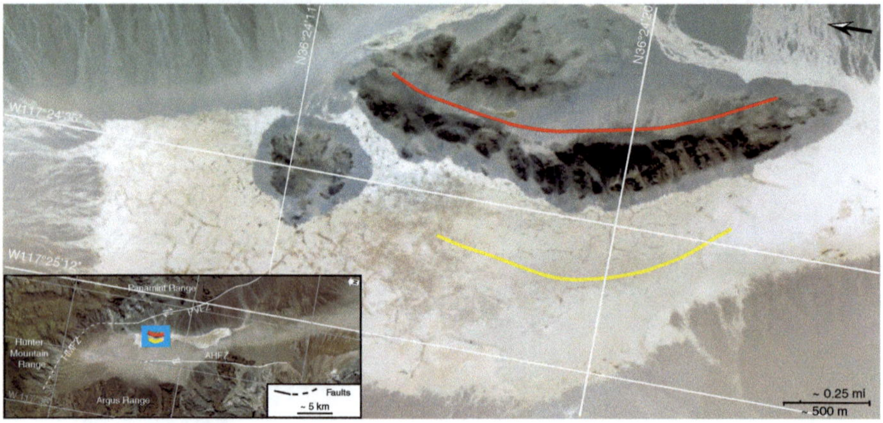

Fig. 4.8 Pre-quaternary fault sketched in *red* (after [22]. Modern mega-crack (~1.6 km/1 miles length) with similar shape on satellite image of Panamint valley. Image © 2010 Google and DigitalGlobe

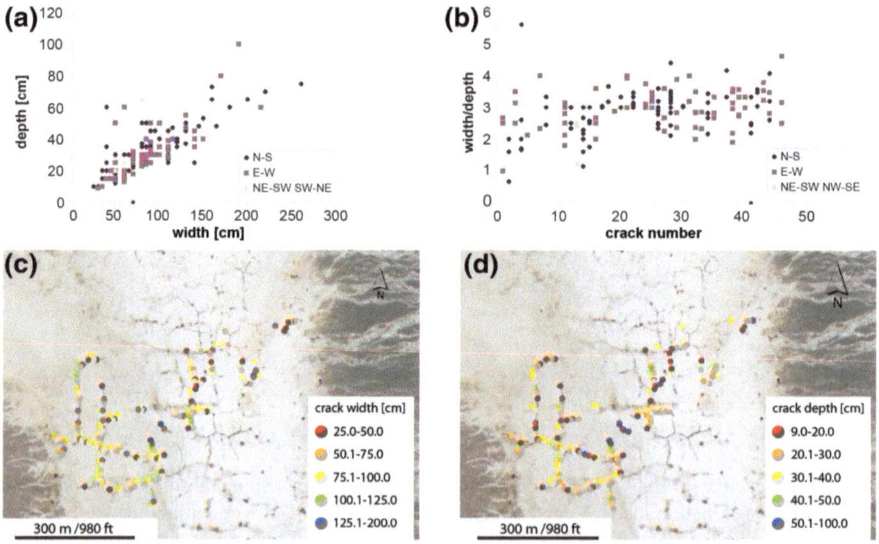

Fig. 4.9 **a** Crack width versus depth related to strike of crack. **b** Crack # versus width/depth related to strike of crack. **c/d** Crack width (*left*) and crack depth (*right*) colour coded. Area of measurements marked in Fig. 4.3. In general, crack width and depth are positively correlated. In all orientations width/depth ratio does not appear to have any correlation to orientation. Satellite images © 2010 Google and DigitalGlobe

composition of the crack walls and the soft material within the cracks. The main components are quartz, calcite, halite, illite/muscovite, feldspar and dolomite. Furthermore, chlorite, kaolinite, haematite, thenardite and the amphiboles actinolite, hornblende and riebeckite are present as minor constituents. In the case of a

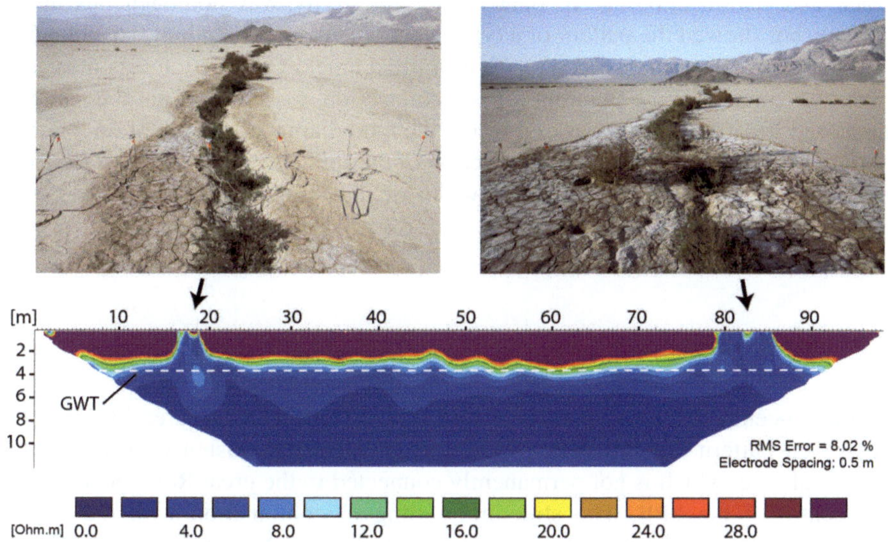

Fig. 4.10 Characteristic ground resistivity profile over two mega-cracks: lower resistivity of cracks compared to the surrounding recognizable until at least 3.5 m (~11.5 ft) depth where the *GWT* groundwater table is located. Continuation below the current *GWT* is likely, but can not be determined due to small differences in resistivity as a result of the high electric conductivity of water

filled fissure, gypsum and anhydrite are present as accessories. The dominant minerals recognized in the samples of the Panamint Dunes are quartz, feldspar and calcite. Additionally, illite/muscovite, chlorite, kaolinite and amphiboles are found. The presence of muscovite and feldspar that are largely affected by chemical weathering during exposure indicate a location close to the source area. Additionally, heavy minerals that are deposited very early during sediment transport can be found throughout the dune field. Swelling clays of the vermiculite and smectite groups were found in the filtered clay fraction.

With the intention to gain information about the influence area, the shape and the vertical extent of the crack's thirteen ground resistivity profiles were measured perpendicular to the cracks on the playa surface (Figs. 4.3, 4.10). The profiles were taken in the northern and southern part of North Panamint Playa to identify local differences (e.g. a varying groundwater table). In most of the cases, the sediments in the cracks were still wet, even 3 weeks after the last rain, prior to the measurements. As a result of the water content and its salinity, the measured resistivity values were very low. The groundwater table was identified without notable local differences at a depth of ~3.5 m (11.5 ft) by its low resistivity (Fig. 4.10). Below that depth the resistivity differences decrease as a result of the high water content and high salinity. The influence area is characterized by higher water content and higher salt content, developed due to enhanced evaporation. This is indicated by low resistivity values in shallow depth (<3.5 m/11.5 ft) on all profiles. In some profiles, no differences in resistivity were noticed between the sediment below a crack and its surrounding. In other cases, resistivity was homogeneous at the surface, but showed the pattern of a typical crack between a depth of 0.5 and 3.5 m (1.6 and 11.5 ft), probably as a result of a filled crack. One section was measured over two mud diapirs, without direct connection to a crack on the surface. Increased resistivity was measured below the mud diapirs at a depth of up to 3.5 m (11.5 ft). In contrast to the visual detection of the influence area, ground resistivity measurements revealed a maximum width of the influence area of 12 m at the playa surface.

4.5 Discussion

The depo-centre interpreted at the centre of the German study area [38] and the polygonal pattern identified on TWT-slices indicate the existence of an isolated perennial lake, which is not permanently connected to the great Rotliegend lake in the centre of the SPB. The structural setting in combination with the similar depositional environment and sedimentary facies that are show in Fig. 4.5 clearly identify the Panamint Valley in California as a suitable modern analogue for the graben/halfgraben setting of the study area at the south-western margin of the SPB. A fluvial origin of the lineaments in the seismic data can be ruled out. Geometries on seismic time slices, horizon slices or vertical sections do not indicate a fluvial origin of these features as it was recognized e.g. in the overlying Triassic Buntsandstein or Keuperian. Furthermore, the small topographic relief at the centre of the German study area during the development of these lineaments [38] and the low energetic fluvial deposits observed in the cores of the wells support this conclusion.

4.5.1 Mega-Crack Development

Hypotheses about the development of megacracks in playas of the western United States involve variations of the groundwater table piezometric surfaces, capillary effects and soil moisture [28]. Furthermore, fissured playas seem to possess significantly higher amounts of clay minerals compared to non-fissured areas [29]. Messina et al. [26] discuss synaeresis as a possible reason for crack evolution, caused by swelling and shrinking of clay minerals due to wetting and drying cycles. The influence of tectonics on cracking processes has to be taken into consideration in tectonically active regions such as the Basin and Range Province. Huge earthquake-induced cracks in ruptured playa sediments on Lavic Lake in California, United States are discussed by Treiman et al. [37]. It is likely that the major boundaries of the mega-crack pattern on Panamint Playa are provided by the active faults. The dominant E–W and N–S orientation of the mega-cracks shown in the rose diagram (Fig. 4.11) support this assumption.

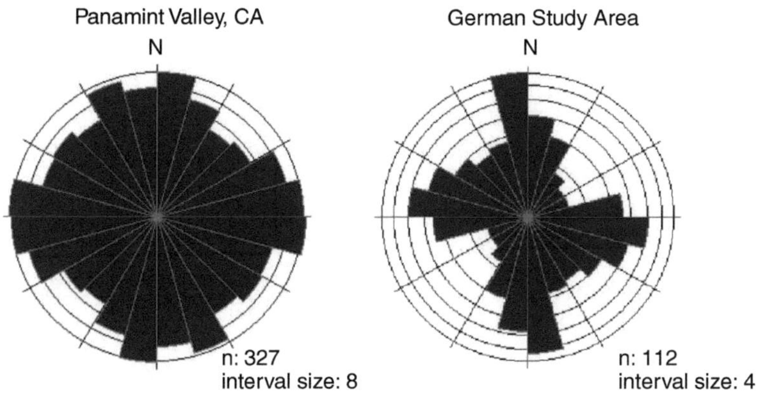

Fig. 4.11 Rose diagrams comparing the orientations of the mega-cracks on Panamint playa, California to faults and lineaments on seismic TWT-slices from Northern Germany

Cartwright et al. [4] proposed the hypothesis of a "progradation sequence for the development of polygonal faults in alternating depositional units that favour or preclude synaeresis during shallow burial", into a polygonal fault system during time. The model relates the development of a polygonal fault system to three successive steps. Heterogeneity in sedimentary layers is established by synaeresis processes in the subsurface. The differences in competence in the layers then result in the development of weakness zones during compaction. These zones serve as preferred location for fault development and predict the framework for the evolution of a polygonal fault system.

Four stages of what might represent a crack evolution cycle (Fig. 4.12) are postulated on the basis of the studies at the Panamint dry lake. Stage one can be considered as the subsurface phase of crack development. No signs of a cavity are visible at the surface until the surface finally collapses. After cracking and fracturing of the surface, the typical crack shape develops and vegetation starts growing in and/or along the cracks in stage two. During stage three, cracks are subsequently refilled, predominantly by wind-blown sand and collapsing crack-walls. Vegetation associated with the cracks serves as sediment traps. In contrast to the remaining playa surface, the crack-filling is composed of coarser grained material ranging from clay to coarse grained sand. In some cases sand and silt bulges form around bushes situated at filled cracks. In the final phase of crack evolution these reliefs get eroded, so that the pre-existing crack can only be identified by colour differences compared to the surrounding playa surface. During lake flooding, a new layer of clay is deposited, which erases the surface

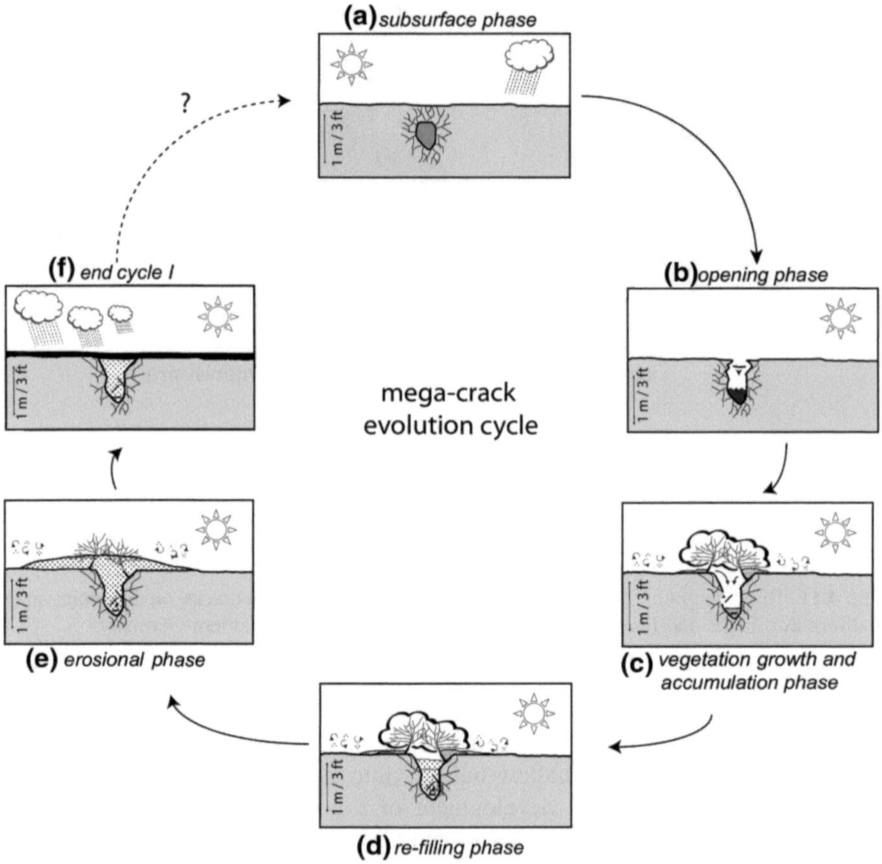

Fig. 4.12 Mega-crack evolution cycle. See text for detailed description and Fig. 4.7 for example photos of crack cycle phases: Subsurface phase—Fig. 4.7a, opening phase—Fig. 4.7b, vegetation growth and accumulation phase—Fig. 4.7d, re-filling phase—Fig. 4.7e, erosional phase—Fig. 4.7f, end of the crack cycle—Fig. 4.7g

morphology of the underlying crack-fill. A new cycle of cracking starts by the drying period following the next rainfall. The stable position of the mega-cracks on the playa surface is consistent with our hypothesis of the evolution of the mega-crack system. A new crack cycle preferably develops at a location of a refilled crack due to the heterogeneity of the sediment fill of the cracks in comparison to the surrounding clay dominated homogeneous playa surface. Observations during the field study suggest that the timeframe for the refilling phase of the cracks through collapsing crack-walls and wind-blown sand is potentially limited to a few years. After 2 hours of heavy winds the fresh cracks were already filled by 10 cm ($\sim$4 in.) of wind-blown sand. This leads to the assumption that during ongoing sedimentation and multiple crack development cycles, the vertical extent of the cracks successively increases.

Ground resistivity measurements show that variations in porosity and water content due to mega-crack development outreach the area recognizable by visual inspection in depth. Cracks connect to the groundwater table, which is located ~3.5 m (11.5 ft) below the surface of the playa (Fig. 4.10). This connection can be explained by the existence of a fractured zone, not visible from the surface below the broken-in part of the crack. The saline content in combination with the clay dominated playa surface leads to the development of a dense surface with very limited drainage opportunities. After a rain or flooding event, surface water evaporates or migrates into the depressions of the mega-cracks. Although the crack-fill is dominated by wind-blown sand with good porosities and loose packing that commonly enhance drainage, the cracks remain wet for a certain period of time after the remaining playa surface has dried. This can be explained by the fact that the collapsed playa surface and crack walls partly seal the crack and thus slow down the drainage process. The water gathered in the depressions formed by the cracks enables vegetation growth throughout the year. These circumstances explain the lower resistivity below the cracks compared to the higher resistivity values of the surrounding mudflat.

Refilled cracks that cannot be identified on the playa surface anymore could be imaged in ground resistivity measurements. In most of the cases, they still show the lower resistivity pattern compared to the surrounding, suggesting that they are still moist.

A previous study by Langford and Eshete [23] in the Chihuahuan Desert in Texas, United States recognized the influence of mega-cracks at depths of up to 8 m. Considering that the development of the huge cracks is associated with volumetric changes of clay minerals, it is likely that these processes represent a direct link to fluctuations of the groundwater table. Consequently, the cracks could not exceed below the depth of the groundwater table without a tectonic component. Furthermore, abrupt mechanical stress build-up is seen as a major trigger for clay liquefaction and dispersion [33], which is essential for mud diapir development.

The impact of tectonics on the evolution and development of megacrack systems is difficult to constrain. Depth and width of 34 open cracks were measured and analysed, considering their individual orientations. The same routine was applied to the influence area with a total of 112 open and filled cracks being analysed. Results do not show any preferred directions, allowing no immediate conclusion about a genetic link between crack development and the regional fault system. However, the shape and orientation of the longest (filled) crack on North Panamint Playa perfectly matches the trace of a pre-quaternary fault at the Panamint Buttes (Fig. 4.8). Furthermore, the association of at least 14 pairs of mud volcanoes with crack systems implies a certain involvement of tectonics as trigger mechanism for mud fluidization. Explaining mud volcano activity solely by contraction and expansion of clay minerals during multiple wetting and drying cycles seems unlikely. High sedimentation rates, which would rapidly increase the load of the overburden and lead to liquefaction of clays in the subsurface, can be ruled out for the Panamint Playa. Swelling clays of the vermiculite and smectite groups were identified in the filtered clay fraction of the samples, but muscovite and illite which have much less swelling potential [10] were also recognized on the playa surface. Signs of mud diapirs on the playa surface close to the newly developed cracks were not observed. It is likely that the combining effects of tectonics and volume variation in clay minerals force the crack development. According to Dimitrov [9], the occurrence of mud volcanoes is frequently controlled by recent tectonic activity, particular compressional, and the presence of thick, fine-grained, soft, plastic sediments deep in the sedimentary succession. Driving forces can be structurally or tectonically induced high pore-fluid pressures. The final trigger to mobilize the muds is then provided by rapid tectonic stress build-up through seismic activity. However, Messina et al. [26] invoke the hydraulic head pressure of the surrounding alluvial fans as a possible driving force for the formation of mud volcanoes or spring mounds in the Panamint Valley.

4.5.2 Fault Development

Frequent fluid circulation enhanced by fractures and sandy fissure-fills are the reason for the heterogeneous composition of cracked areas, in contrast to the homogenous surrounding playa surface comprising clay and fine silt. Although a direct link between orientation of cracks and the fault system in Panamint Valley was not determined with the measurements, the possibility can not be ruled out. Satellite images and rose plots give a hint that the orientation of some cracks is influenced by the local stress field. Tectonic processes are also known as a trigger for clay liquefaction and the development of mud diapirs. However, a preferred development of mud diapirs associated with cracks of a distinct orientation was not noticed. The contrast in competence of cracked areas to their surrounding makes them a preferred location concerning stress release [4]. This hypothesis is not only valid for the time of the dry lake being located at the surface but also after the sedimentation of the overburden, due to the difference in compaction between inhomogeneous former cracked areas to the more or less homogeneous surrounding of formerly non-cracked areas. The increased stress and the different compaction caused by the overlying sediments may first result in the development of weakness zones. In addition, the better porosity and permeability, as a result of the higher percentage of sand in the crack-fills, represents preferred pathways for fluid migration. With time and burial depth, some of these heterogeneous zones can develop into faults or even into a polygonal fault system as described by Cartwright et al. [4]. It is likely that only zones with a vertical influence of multiple metres develop into faults. Consequently, repeated crack development cycles at the particular locations are a basic requirement.

In the German study area, the faults that developed along the predominant N–S and E–W stress field (Fig. 4.11) show the highest offsets. Similar to the assumption that the major framework for mega-cracks on Panamint Playa is likely provided by faults, Upper Rotliegend fault activity could have served as structural grain for the further development of weakness zones to faults, and thus explain the different offsets between major and minor faults on today's seismic data dependant on their orientation. In contrast to the rhomboidal fault pattern that is common in the Rotliegend of the SPB [6], our data set includes additional low-offset faults and sub-seismic features with multiple orientations that build a polygonal pattern on the variance map (Figs. 4.2, 4.11). The low offsets along these faults could also indicate to the immature state of a polygonal fault system [4], the deep burial of more than 4.5 km (2.8 miles) and the influence of the adjacent salt dome on the seismic data are a challenge for identifying the preserved pattern.

4.5.3 Implication for Hydrocarbon Migration/Systems and Reservoirs

In arid climates, a close interfingering of different sedimentary facies is a common feature. In desert settings, dunes can directly overly mudflats or pond deposits. Dune intervals from cores from Northern Germany, identified as host rocks in the reservoirs, seldom outreach a thickness of 2.5 m (8.2 ft). Interdune deposits comprising pond, mudflat and wet to damp sandflat sediments with higher clay content and low porosity and permeability are commonly interbedded with dune deposits (Fig. 4.13).

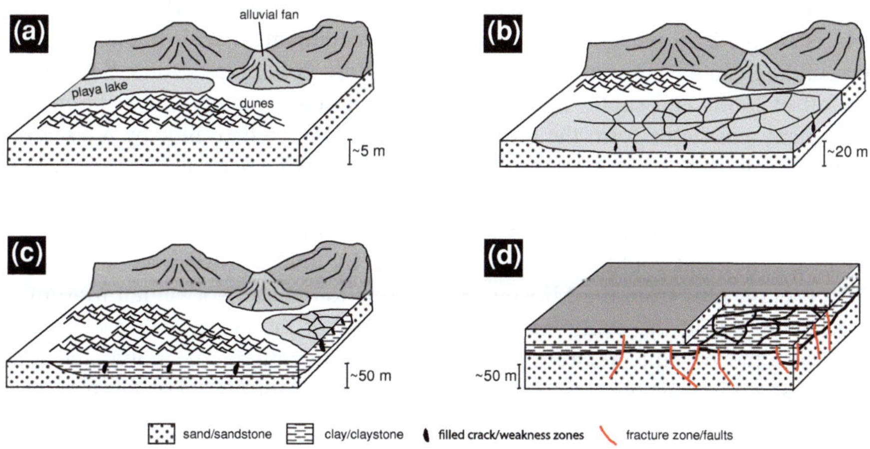

Fig. 4.13 Model for hydrocarbon systems in arid climate settings. Changes in sedimentary facies distribution in arid climates do not necessarily require changes in the sedimentary setting (**a–c**). **b** Mega-cracks system develops on playa surface. **c** Depth influence of cracks increases with the number of cracking cycles (see Fig. 4.10). Location of the playa lake can vary. **d** Faults developed at weakness zones as consequence of competence contrast between crack-fill and remaining playa surface. Weakness zones and faults might provide fluid migration pathways, connecting reservoir intervals (aeolian deposits) separated by sealing lithologies (lake deposits). In the example of the German tight gas field, fluid migration along faults resulted in increased cementation and inversion of porosity and permeability [13, 14] of reservoir rocks. Note different scales

Applying the knowledge about the polygonal pattern of cracks on the surface of dry lakes in the western United States to the Upper Rotliegend II in Northern Germany, I suggest that dune deposits which comprise reservoir intervals separated by impermeable lake and pond deposits could be hydraulically connected by more permeable crack-fills, which developed into weakness zones during burial and finally into faults serving as fluid pathways during times of hydrocarbon migration. However, in the case of the tight gas field in Germany, the fast decrease of initial production rates indicates a compartmentalisation of the reservoir. The decreased porosity and permeability can be explained by an inversion of original porosities and permeabilities as a result of quartz overgrowth and secondary illite crystallization in the proximity of fault zones [13, 14]. Well information reveals that porosity and permeability values are lower proximal to the fault compared to distal locations, where production is carried out economically. Consequently, I conclude that the vertical connection provided by the faults initially promoted fluid flow, enabling fluid migration between aeolian intervals that are vertically separated by lake deposits. During time, this fluid migration resulted in the inversion of the primary good porosity and permeability at the faults and their proximity [13, 14]. Considering the significant intervals of lake/pond deposits in the core data and the information from seismic data, vertical and horizontal compartmentalisation seems likely for the tight gas reservoir in Germany. In the previous study (see Chap. 3) I suggested that only some of the weakness zones influenced by synaeresis developed into faults. Such small-scale faults with minor offsets are hardly resolvable on the basis of available seismic data at a depth of 4.5 km (2.8 miles), where the tight gas field is situated. However, even small offsets accompanied with fracturing in the surrounding can have a significant influence on fluid flow in reservoirs. Such minor, low offset faults and subtle lineaments are described as compartment boundaries from many gas fields form the Groningen area in the Netherlands by Van Hulten [39]. Not all parts of the pattern, identified on TWT-slices using the variance seismic attribute, can be identified as faults in vertical sections (Fig. 4.2). Some areas, where the pattern can be observed on the TWT-slice, are accompanied by a negative relief and could record effects caused by mud diapirism as described by Hustoft et al. [20] from the Miocene Kai Formation at the Norwegian Margin.

Knowing about the evolution and composition of the polygonal cracks on dry lakes and the possible relation to a regional fault pattern, the polygonal crack pattern identified on the basis of the seismic data from Northern Germany was re-examined.

The orientation of the polygonal pattern on the seismic was compared to the prevailing stress orientation of the Upper Rotliegend II in Northern Germany. Faults striking ~N–S, oriented perpendicular to the extensional stress pattern, show higher cumulative vertical offsets than ~E–W oriented faults. Consequently, I assume that preferably the ~N–S oriented existing weakness zones or faults were active over a longer period of time, due to their matching orientation to the regional stress field and higher offsets than the ~E–W oriented faults. This is supported by the observation that the number of lineaments on TWT-slices, identifiable as faults (by a visible offset) in vertical sections is higher in N–S direction compared to E–W direction. However, it does not dismiss E–W running lineaments or the parts of the polygonal pattern visible on the seismic time slices as weakness zones, faults and/or fluid migration pathways.

4.6 Conclusions

Polygonal mega-crack development on dry lakes is a common feature in arid climate settings. After burial, former crack zones develop into weakness zones that are preferred locations for the development of (low offset) faults that provide possible fluid pathways. I conclude that the crack development combined with the mechanisms of the evolution of a polygonal fault system may play an important role in influencing hydrocarbon reservoir compartmentalisation. Reservoir connectivity is influenced by these zones, both in a positive or negative way since fluid reactions may lead to secondary porosity enhancement or pore-destructive cementation. Concerning reservoir quality, the results of this study therefore recommend paying higher attention to the 3D distribution of subtle, small-scale faults and their organisation in possible polygonal networks.

References

1. Antrett P, Vackiner A, Stollhofen H, Kukla P (2009) Sedimentological reservoir characterization of a tight gas field in the Upper Rotliegend II Wustrow and Bahnsen members, Ostfriesland, Germany. In: 6th Annual conference of SEPM-CES sediment 2009, Abstracts and Field Guide, Krakow, Poland, 7, ISBN: 978-83-7538-632-5
2. Brown AR (2001) Data polarity for the interpreter. Lead Edge 20:549
3. Burchfield BC, Hodges KV, Royden LH (1987) Geology of Panamint valley—Saline valley pull-apart system, California: Palinspastic evidence for low-angle geometry of a neogene range bounding fault. J Geophys Res 92 (B10):10,422–10,426
4. Cartwright J, James D, Bolton A (2003) The genesis of polygonal fault systems: a review. In: Van Rensbergen P, Hillis RR, Maltman AJ, Morley CK (eds) Subsurface sediment mobilization. Geological Society, London, Special Publications, vol 216, pp 223–243
5. Davies RJ, Ireland MT, Cartwright JA (2009) Differential compaction due to the irregular topology of a diagenetic reaction boundary: a new mechanism for the formation of polygonal faults. Basin Res 21:354–359
6. de Jager J (2007) Geological development. In: Wong TE, Batjes DAJ, de Jager J (eds) Geology of the Netherlands. Royal Netherlands Academy of Arts and Sciences, pp 5–26
7. Densmore AL, Anderson RS (1997) Tectonic geomorphology of the Ash hill fault, Panamint Valley, California. Basin Res 9:53–63
8. Dewhurst DN, Cartwright JA, Lonergan L (1999) The development of a polygonal fault system by synaeresis of colloidal sediments. Mar Pet Geol 16:793–810
9. Dimitrov LI (2002) Mud volcanoes-the most important pathway for degassing deeply buried sediments. Earth Sci Rev 59:49–76
10. Foster MD (1954) The relation between composition and swelling in clays. Clays Clay Miner 3(1):205–220

11. Gast R (1995) Sequenzstratigraphie. In: Plein E (ed) Stratigraphie von Deutschland I; Norddeutsches Rotliegendbecken—Rotliegend-Monographie Teil II. Courier Forschungs-Institut Senckenberg, Frankfurt a. M, vol 183, pp 47–54
12. Gast R, Pasternak M, Piske J, Rasch H-J (1998) Das Rotliegend im nordostdeutschen Raum: Regionale Übersicht, Stratigraphie, Fazies und Diagenese. Geologisches Jahrbuch A 149:59–79
13. Gaupp R, Solms M (2005) Sedimentological and petrological investigations. In: Paleo oil—and gasfields in the Rotliegend of the North German Basin: effects upon hydrocarbon reservoir quality (Paläo-Öl- und Gasfelder im Rotliegenden des Norddeutschen Beckens: Wirkungen der KW-Migration auf die Speicherqualitäts-Entwicklung.) DGMK Research Report, 593–598: tight gas reservoirs—natural gas for the future, pp 1–1, 1–44
14. Gaupp R, Matter A, Platt J, Ramseyer K, Walzebuck JP (1993) Diagenesis and fluid evolution in deeply buried Permian (Rotliegende) gas reservoirs, Northwest Germany. AAPG Bull 77(7):1111–1128
15. George GT, Berry JK (1993) A new palaeogeographic and depositional model for the Upper Rotliegend of the UK sector of the Southern North sea. In: North CP, Prosser DJ (eds) Characterization of fluvial and aeolian reservoirs. Geological Society, London, Special Publications, vol 73, pp 291–319
16. Glennie KW (1972) Permian Rotliegendes of Northwest Europe interpreted in light of modern desert sedimentation studies. AAPG Bulletin 56(6):1048–1071
17. Glennie KW (1983) Early Permian (Rotliegendes) palaeowinds of the North sea. Sed Geol 34:245–265
18. Glennie KW, Higham J, Stemmerik L (2003) Permian. In: The millennium atlas; petroleum geology of the central and northern North Sea. Geological Society of London, London, United Kingdom, pp 91–103
19. Hodges KV, McKenna LW, Stock J, Knapp J, Page L, Sternlof K, Silverberg D, Wüst G, Walker JD (1989) Evolution of extensional basins and basin and range topography west of Death valley, California. Tectonics 8:435–468
20. Hustoft S, Mintert J, Bünz S, Nouzé H (2007) High-resolution 3D-seismic data indicate focussed fluid migration pathways above polygonal fault systems of the mid-Norwegian margin. Mar Geol 245:89–106
21. Jayko AS, Forester RM, Kaufman DS, Philips FM, Yount JC, McGeehin J, Mahan SA (2008) Late Pleistocene lakes and wetlands, Panamint valley, Inyo county, California. Geol Soc Am 439:151–184 Special Paper
22. Jennings CW, Bryant WA, Saucedo G (2010) Fault activity map of California: California geological survey 150th anniversary. California geologic data map series, Map No. 6, scale 1:750,000
23. Langford RP, Eshete T (2000) Fissure behaviour in the Chihuahuan desert, and depth estimation using sediment strength. Env Eng Geosci VI 4:301–309
24. Legler B (2005) Faziesentwicklung im Südlichen Permbecken in Abhängigkeit von Tektonik, eustatischen Meeresspiegelschwankungen des Proto-Atlantiks und Klimavariabilität (Oberrotliegend, Nordwesteuropa). Schriftenreihe der Deutschen Gesellschaft für Geowissenschaften, 47
25. Mazac O, Kelly WE, Landa I (1987) Surface geoelectrics for groundwater pollution and protection studies. J Hydrol 93:277–294
26. Messina P, Stoffer P, Sith WC (2005) Macropolygon morphology, development, and classification on North Panamint and Eureka playas, Death Valley National Park CA. Earth Sci Rev 73:309–322
27. Motts WS, Carpenter D (1968) Report of test drilling on Rogers, coyote, Rosamond, and Panamint playas in 1966. In: Neal JT (ed) Playa surface morphology: miscellaneous investigations. office of aerospace research, U.S. Air Force. AFCRL-68-0133, pp 31–57
28. Neal JT, Motts WS (1967) Recent geomorphic changes in playas of the western United States. J Geol 75:511–525

29. Neal JT, Langer AM, Kerr PF (1968) Giant desiccation polygons of great basin playas. Geol Soc Am Bull 79(1):69–90
30. Plein E (1995) Stratigraphie von Deutschland I: Norddeutsches Rotliegendbecken, Rotliegend-Monographie Teil II, Courier Forschungs-Institut Senckenberg, Frankfurt a. M., 183
31. Reheis MC, Sawyer TL (1997) Late Cenozoic history and slip rates of the fish lake valley, emigrant peak, and deep springs fault zones, Nevada and California. Geol Soc Am Bull 109:280–299
32. Rieke H (2001) Sedimentologie, Faziesarchitektur und Faziesentwicklung des kontinentalen Rotliegenden im Nordostdeutschen Becken (NEDB). Ph.D. thesis, Potsdam University, Germany
33. Rowell DL, Payne D, Ahmad N (1969) The effect of the concentration and movement of solutions on the swelling, dispersion, and movement of clay in saline and alkali soils. J Soil Sci 20(1):176–188
34. Smith RSU (1976) Late quaternary pluvial and tectonic history of Panamint valley, Inyo and San Bernardino Counties, California. Published dissertation, California Institute of Technology, University Microfilms, Ann Arbor, Michigan
35. Stollhofen H, Bachmann GH, Barnasch J, Bayer U, Beutler G, Franz M, Kästner M, Legler B, Mutterlose J, Radies D (2008) Upper Rotliegend to early cretaceous basin development. In: Littke R, Bayer U, Gajewski D, Nelskamp S (eds) Dynamics of complex intracontinental basins. The central European basin system. Springer, Berlin, Heidelberg, pp 181–210
36. Strömbäck AC, Howell JA (2002) Predicting distribution of remobilized aeolian facies using sub-surface data: the Weissliegend of the UK Southern North sea. Pet Geosci 8:237–249
37. Treiman JA, Kendrick KJ, Bryant WA, Rockwell TK, McGill SF (2002) Primary surface rupture associated with the Mw 7.1 16 October 1999 hector mine earthquake, San Bernardino County, California. Bull Seismol Soc Am 92(4):1171–1191
38. Vackiner AA (2011) Sedimentary facies reconstruction and kinematic restoration of an upper Permian tight gas field, north-western Germany. PhD thesis, RWTH Aachen University, Germany
39. Van Hulten FFN (2010) Geological factors effecting compartmentalization of Rotliegend gas fields in the Netherlands. In: Jolley SJ, Fisher QJ, Ainsworth RB, Vrolijk PJ, Delisle S (eds) Reservoir compartmentalization: special publication 347. Geological Society, London, pp 301–315
40. Zhang P, Ellis M, Slemmons DB, Mao F (1990) Right lateral displacements and the Holocene slip rate associated with prehistoric earthquakes along the southern Panamint valley fault zone. J Geophys Res 95(84):4857–4872
41. Zhou W, Beck BF, Stephenson JB (2000) Reliability of dipole–dipole electrical resistivity tomography for defining depth to bedrock in covered karst terrancs. Environ Geol 39(7):760–766
42. Ziegler PA (1982) Geological atlas of western and Central Europe. Shell Internat, Petrol Maatsch, The Hague
43. Ziegler PA (1990) Collision related intra-plate compression deformations in Western and Central Europe. J Geodyn 11:357–388

Chapter 5
Facies Analysis of Cores from 4 Wells from Northern Central Germany

5.1 Introduction

In this study, 168 m core material from 4 wells (Fig. 5.1) of the Upper Rotliegend II in the SPB in northern central Germany was analysed with focus on sedimentary facies and facies distribution. The study area is located $\sim$200 km east of the East Frisian tight gas field (Fig. 5.1; see Chaps. 3 and 4). Sedimentation during the Upper Rotliegend II in the SPB was controlled by arid to semi-arid climate, and deposition was dominated by siliciclastics and subordinate evaporites [6]. George and Berry [5], Strömbäck and Howell [12] and Legler [8] suggested ephemeral fluvial (wadi), aeolian, sabkha, and lacustrine environments as the four major facies associations. Huge amounts of aeolian sediments were provided by strong winds from the east [4, 10]. While the overall geological setting in the two study areas is similar (see Sect. 1.4), regional differences in the depositional setting occur due to the location of the second study area close to the centre of the SPB. Lake level changes of the perennial saline lake in the central part of the SPB during the Dethlingen [3] strongly influence the sedimentation in northern central Germany. In contrast to the East Frisian tight gas field, where core analysis was carried out in the Wustrow and Bahnsen Members, the investigated intervals of the second interest area included younger strata from the Bahnsen to Munster. A comparison of the two study areas leads to a deeper understanding of the regional changes of the depositional environments, the palaeogeography and reservoir rock distribution of the SPB during the Upper Rotliegend II.

P. Antrett, *Characterization of an Upper Permian Tight Gas Reservoir*, Springer Theses, DOI: 10.1007/978-3-642-36294-1_5,

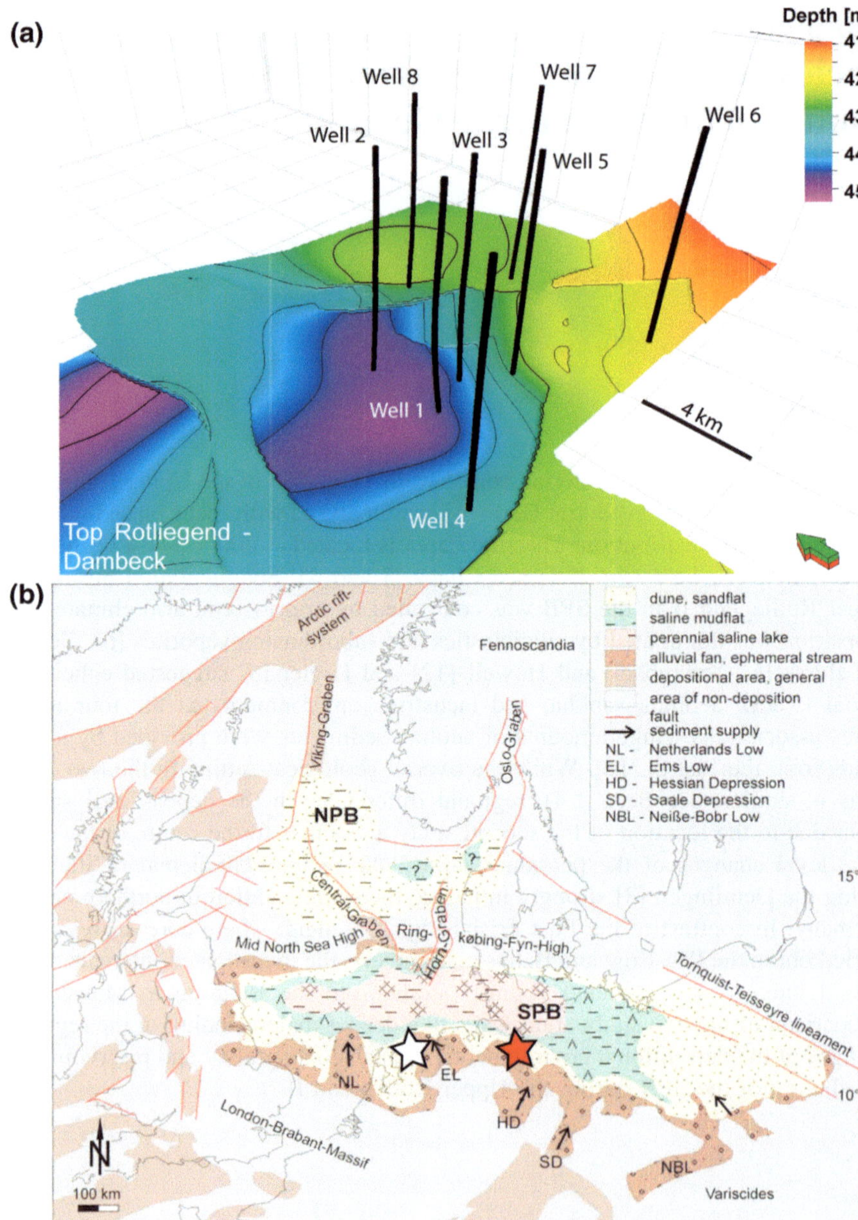

Fig. 5.1 **a** Overview of the wells from the northern central German study area. Wells *1–4* are analysed in this chapter. Wells *5–8* are discussed by Vackiner [13]. Top Rotliegend (Dambeck) horizon displayed. **b** Palaeogeography and general facies distribution of the Upper *Rotliegend* II Elbe Subgroup (compiled from [7, 11, 14]). *NPB* Northern permian basin, *SPB* Southern permian basin. The *white star* marks the East Frisian tight gas field. The *red star* marks the northern central German study area

List of wells

Well	Core	Core metre (m)	Formation	Member
Well 1	Core 1	0.0–9.0	Hannover	n/a
Well 2	Core 1	0.0–13.0	Hannover	Bahnsen
	Core 2	13.0–31.0	Hannover	n/a
	Core 3	31.0–43.3	Hannover	n/a
Well 3	Core 1	0.0–18.0	Hannover	Munster
	Core 2	18.0–36.0	Hannover	n/a
	Core 3	36.0–54.0	Hannover	n/a
	Core 4	54.0–67.8	Hannover	n/a
Well 4	Core 1	0.0–18.0	Hannover	n/a
	Core 2	18.0–29.5	Hannover	n/a
	Core 3	29.5–38.3	Hannover	n/a
	Core 4	38.3–48.3	Hannover	n/a

5.2 Sedimentary Facies in the Cores

5.2.1 Pond/Lake

Pond/lake deposits consist of 100 % structure-less to laminated red, minor green to grey clay. Their occurrence is associated with interdune as well as with ephemeral dry lake settings. Salt efflorescence is often recognized along uncemented fractures in these sediments, which are mainly oriented horizontally, parallel to the bedding. Occasionally, vertically oriented fractures occur (Fig. 5.2).

Fig. 5.2 Characteristic pond/lake deposit

Fig. 5.3 Structureless sandstone lens in a lake margin deposit

5.2.2 Lake Margin

Sediments from lake margin deposition are characterized by irregular cm-scale intercalations of sandstones and clay. Sandstone lenses are poorly sorted, structureless and very fine to fine grained (Fig. 5.3). Ripple lamination is common. While the lake margin deposits are associated with sub-aquatic milieus, intercalation of aeolian sediments into the clay intervals most likely results from dune sets prograding into the standing water body.

5.2.3 Aeolian Mudflat

Aeolian mudflat deposits are wavy laminated claystones with intercalated lenses of siltstone and sandstone. They are characterized by a clay content >50 % Amthor and Ockerman [1]. Their deposition took place in a mainly sub-aquatic milieu without recognisable currents. Desiccation cracks and soft clasts indicate periodical, subaerial exposition of the mudflats. Adhesive sands and silts were synsedimentary blown into the clay deposits (Fig. 5.4). In the cores, the majority of

Fig. 5.4 Characteristic aeolian mudflat succession

Fig. 5.5 Sand injection in an aeolian mudflat from Well 3

mudflat deposits is not laminated, but contain dewatering and desalinization structures. Additionally, post-sedimentary structures such as convolute bedding, flame structures, sand diapirs, and injection are common (Fig. 5.5). In many cases, salt efflorescence occurs along horizontal, bedding-parallel, small, uncemented fracture systems, while vertically oriented fractures are rare.

5.2.4 Damp Sandflat

Damp sandflat deposits are fine grained to medium grained sandstones with minor siltstone content. Clay contents are <20 % [1]. They are characterized by wavy, laminated bedding and contain dewatering and desalinization structures. Post-sedimentary structures, such as convolute bedding, flame structures, sand diapirs, and injections are common (Fig. 5.6). Small current ripples observed in cores are of sedimentary origin. The wet sandflat environments most likely developed during periods of shallow groundwater levels or ephemeral flooding. In some cases, embedded grain flow deposits indicate dry, aeolian depositional areas in the closer vicinity of the wet sandflats (Fig. 5.7). Adhesive sands and minor deflation lags are common due to a temporarily increased moisture content at the sediment surface.

5.2.5 Wet Sandflat

Wet sandflat deposits are very fine to fine grained, poorly sorted sandstones and siltstones with clay contents of 20–50 % [1]. They are characterized by a wavy, irregular bedding (Fig. 5.8). Sand lenses are often embedded in finer grained sediments. They also contain post-sedimentary dewatering structures (Fig. 5.9). The primary sedimentary layering is further disturbed by precipitation and dissolution of salt close to the sediment surface. Small current ripples developed synsedimentarily. The wet sandflats developed at high groundwater levels, close to the sediment surface. Some grain flow deposits indicate an adjacency to dry, aeolian depositional areas. Adhesive sands and minor deflation lags are common due to a temporarily increased moisture content at the sediment surface.

5.2.6 Dry Sandflat

Dry sandflat deposits are characterized by mature, (very) well bedded, and homogeneous, fine to coarse grained sandstones (Fig. 5.10). Characteristic features are a bimodal grain size distribution and cross stratification with dip angles of less than 5° (Fig. 5.11; [9]). Dry sandflats are either associated with dune fields or isolated features in areas of high wind velocities.

Fig. 5.6 Wavy, laminated bedding in a damp sandflat

Fig. 5.7 Damp sandflat from Well 2. Coarse grained grain flow deposit in the centre of the image

Fig. 5.8 Irregular bedding of a wet sandflat from Well 3

Fig. 5.9 Sand lenses interbedded in finer grained material in the *upper part*, and post-sedimentary dewatering in the wet sandflat

Fig. 5.10 Homogeneous dry sandflat from Well 3

Fig. 5.11 Bimodal grain size distribution in a dry sandflat

5.2.7 Low Energetic Fluvial Deposits

Fluvial deposits are composed of low energetic, shallow, braided stream channel systems and sheetflood sediments. Grain sizes vary between clay (predominantly) and coarse grained sand. Furthermore, fluvial reworked sediments are characterized by small rip-up clasts and clay cullets (Fig. 5.12). Small-scale tangential epsilon cross-stratification is a common feature (Fig. 5.13). In general, the lower parts of the small scale cross-beds are characterized by fine grained sediments, while the upper parts show comparative coarser grained sediments. Meandering fluvial channel systems are further identified by fining upward trends.

5.2.8 Aeolian Dune (base)

Aeolian dune deposits are (very) well bedded, fine to coarse grained, mature sandstones. They are characterized by semi-tabular, uni-modally dipping cross-beds with a bimodal grain size distribution. Coarse grained grain flow deposits are closely intercalated with fine grained suspension fallout deposits (Fig. 5.14; [2]). Grain size variations in the sediments may depend on total wind velocities, composition of the source area, and the location of the dunes in the dune field. The aeolian dunes are typically characterized by upward increasing dip angles: their bases consist of sandstones with dip angles of 5–15°, increasing upward to more than 15° [5].

5.3 Core Description

5.3.1 Well 1 (Core 1: 0.0–9.0 m; Fig. 5.15)

The core is dominated by aeolian deposits. Fluvial sediments at the base are overlain by homogenous sandflat deposits (A). The dominance of damp sandflats indicates (B) that aeolian successions developed in areas with relatively shallow groundwater levels. Pond (C) and mudflat deposits (D) are interpreted to originate from interdune areas.

5.3.2 Well 2 (Core 1: 0.0–13.0 m, core 2: 13.0–31.0 m, core 3: 31.0–43.4 m, Figs. 5.16 and 5.17)

The Bahnsen Member (core 3) of Well 2 is characterized by aeolian successions. Sandflats were deposited during periods of shallow groundwater tables. Abundant grain flow input into temporarily wet, damp, and dry sandflats imply that

Fig. 5.12 Fluvial reworked sediments in Well 2

Fig. 5.13 Epsilon cross stratification in low energetic fluvial deposits from Well 3

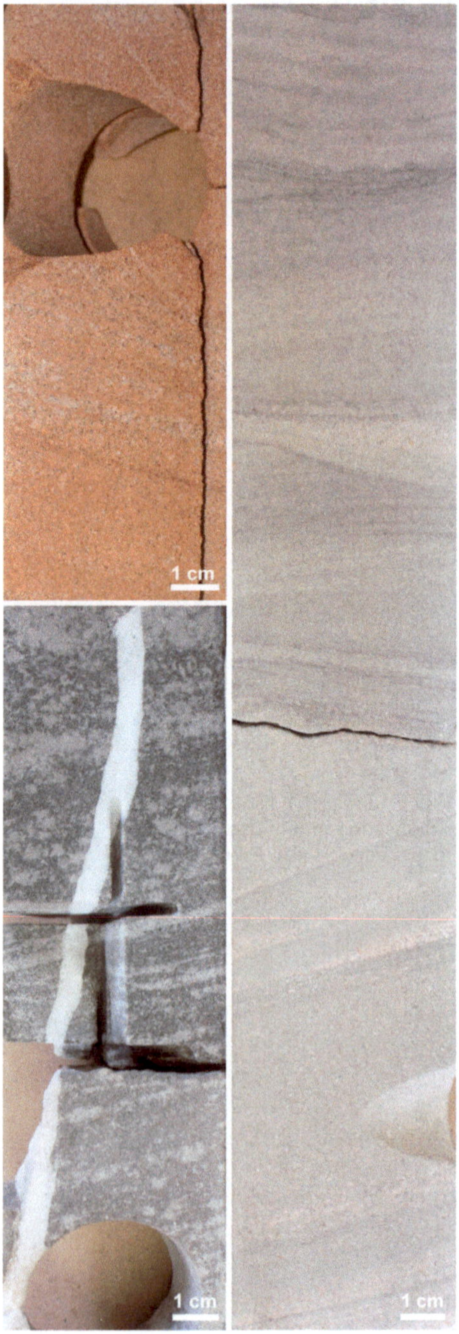

Fig. 5.14 Examples of characteristic aeolian dune (base) successions, including different dune sets, varying dip angles and coarse and fine grained grain flows and suspension fallout

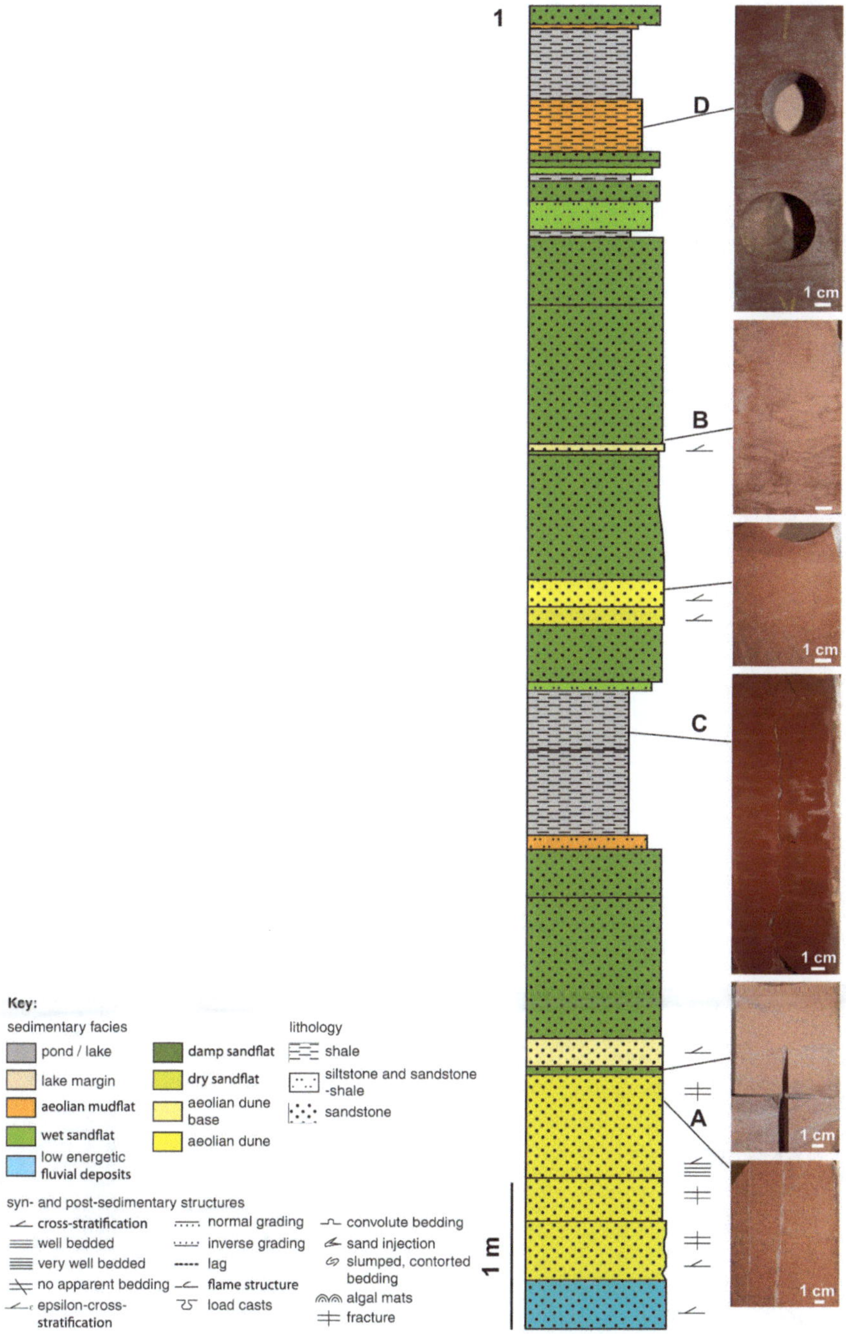

Fig. 5.15 Core *1*, Well 1. Photos of core successions plotted next to the lithology and depositional environment column. *A–D* mark photos of the sedimentary facies referred to in the text

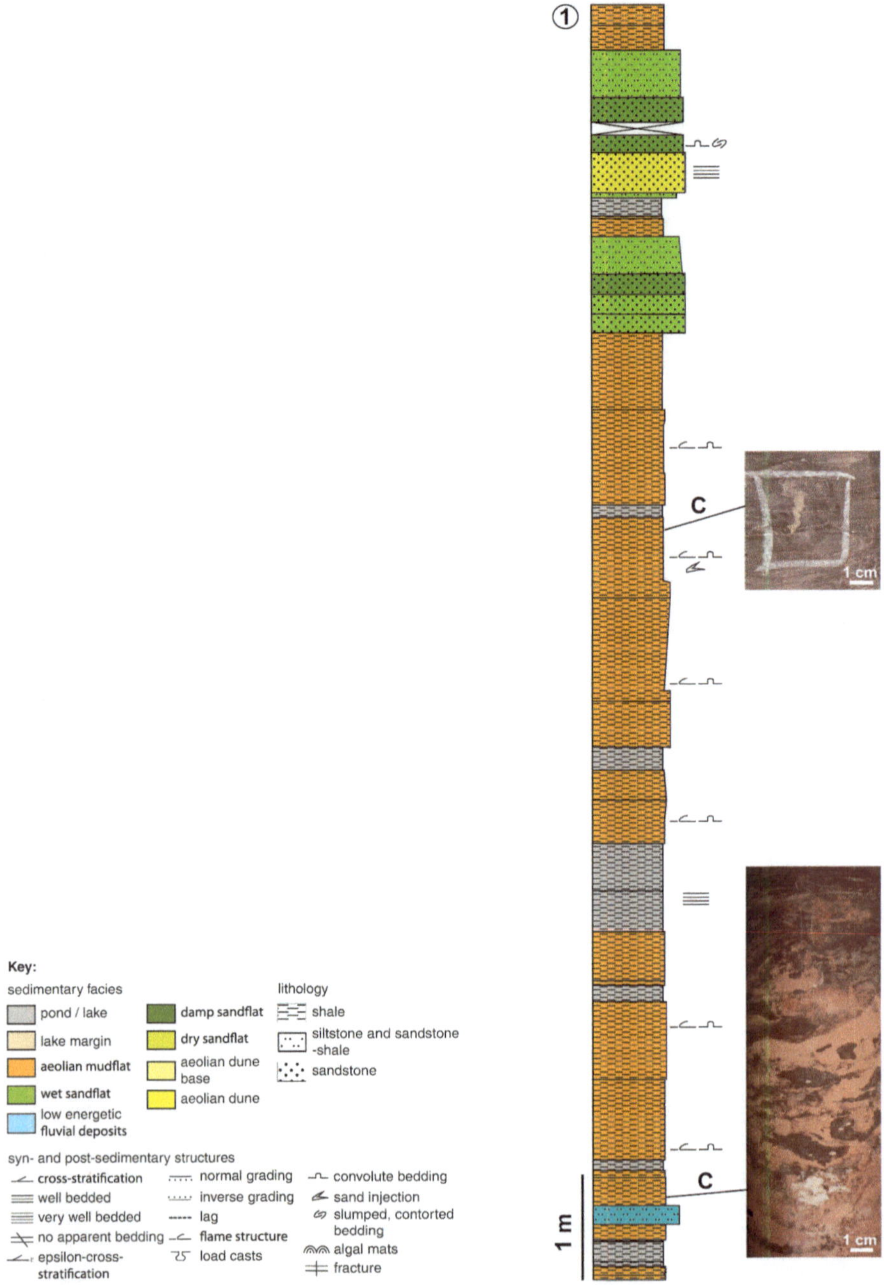

Fig. 5.16 Core *1*, Well 2. Photos of core successions plotted next to the lithology and depositional environment column. *C* marks photos of the features referred to in the text

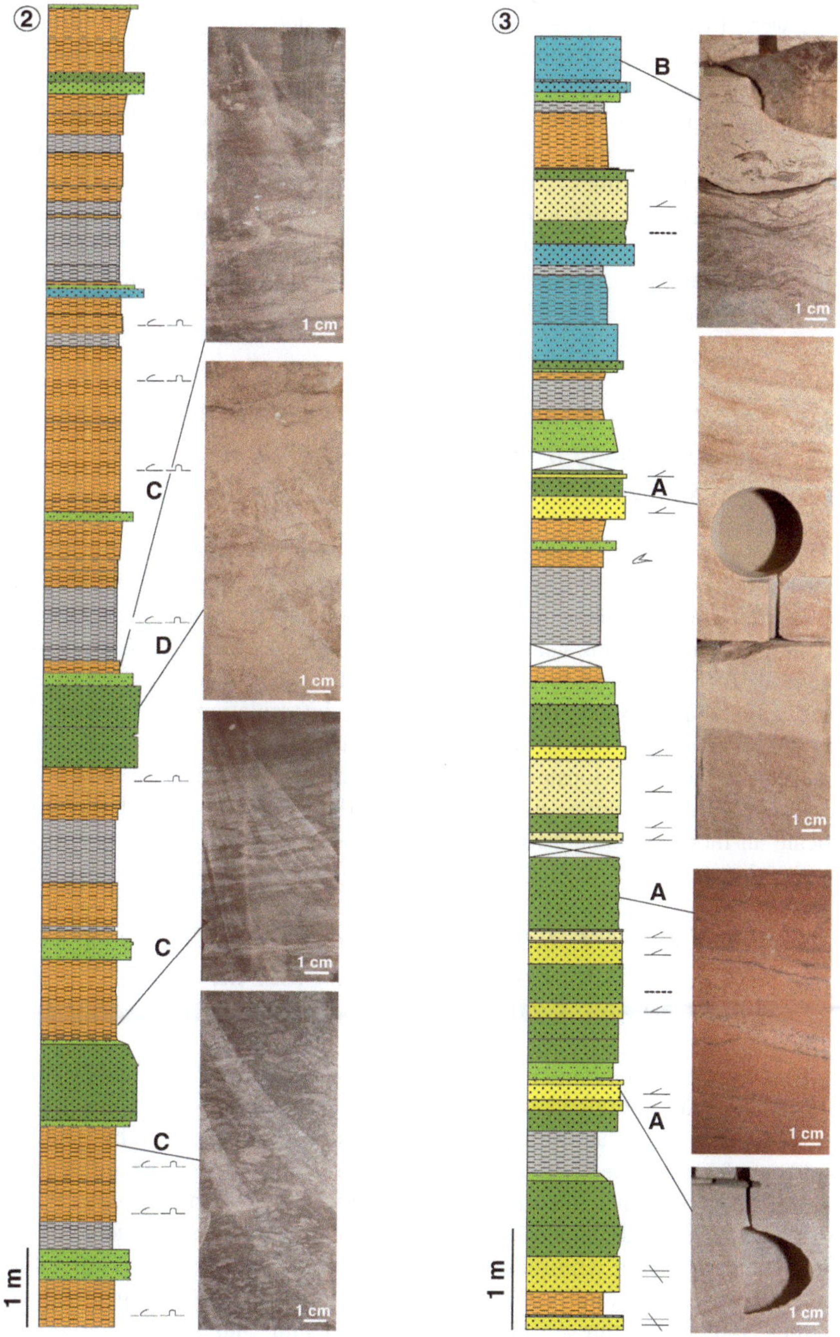

Fig. 5.17 Cores *2* and *3*, Well 2. Photos of core successions plotted next to the lithology and depositional environment column. *A–D* mark photos of features referred to in the text

dunes were located close to the temporarily flooded surface (A). Additionally, fluvial reworked aeolian deposits are interpreted in the upper part of core 3 (B).

Mudflat (C) and sandflat (D) deposits are the dominant sedimentary facies in cores 1 and 2. Deposition took mainly place in a sub-aquatic environment with only ephemeral desiccation. The groundwater table was located close to the sediment surface at all times. Minor fluvial reworked successions likely originate from currents discharging into a standing water body. Multiple damp sandflat deposits and one dry sandflat deposit are interpreted in the Munster Member (D). Mineralization nodules and increased cementation point to an increased diagenetic influence on core 1.

5.3.3 Well 3 (Core 1: 0.0–18.0 m, core 2: 18.0–36.0 m, core 3: 36.0–54.0 m, core 4: 54.0–67.4 m, Figs. 5.18 and 5.19)

Damp sandflats that were deposited in times of shallow groundwater tables represent the dominant sedimentary facies in core 4. During dry weather periods, dry sandflats, aeolian dune bases and aeolian dunes developed (A). Several interdune deposits indicate a lateral shift of the dune system (B). Core 2 and the lower part of core 1 mainly consist of mudflats (C), which were deposited subaquatically. Minor wet and damp sandflat intercalations developed during ephemeral flooding or due to the influence of a shallower groundwater table. The upper part of core 1 (Munster Member) is characterized by a wet to dry aeolian succession (E). Bleaching, mineralisation nodules, and cemented fractures (E) indicate an increased diagenetic overprint of this interval.

5.3.4 Well 4 (Core 1: 0.0–18.0 m, core 2: 18.0–29.5 m, core 3: 29.5–38.3 m, core 4: 38.3–48.3 m, Figs. 5.20 and 5.21)

The lower part of core 4 is dominated by pond/lake and mudflat deposits followed by coarser grained sandstones of a dune/interdune interval in the upper part of the core (A). The lower part of core 3 is also characterized by mudflat and pond/lake deposits. Aeolian sediments that developed under the influence of a shallow groundwater table are present in the upper part. Ephemeral fluvial reworking is likely (B). Cores 2 and 1 are characterized by ephemeral wet deposits that occur intercalated with minor sandflats, mainly deposited under the influence of a varying groundwater table. Post-sedimentary structures such as convolute bedding, flame structures, sand injections (C) and diapirs are common. The uppermost clay-dominated parts of the core show abundant horizontal and minor vertical fracture networks, which might influence gas migration and hydraulic fracturing.

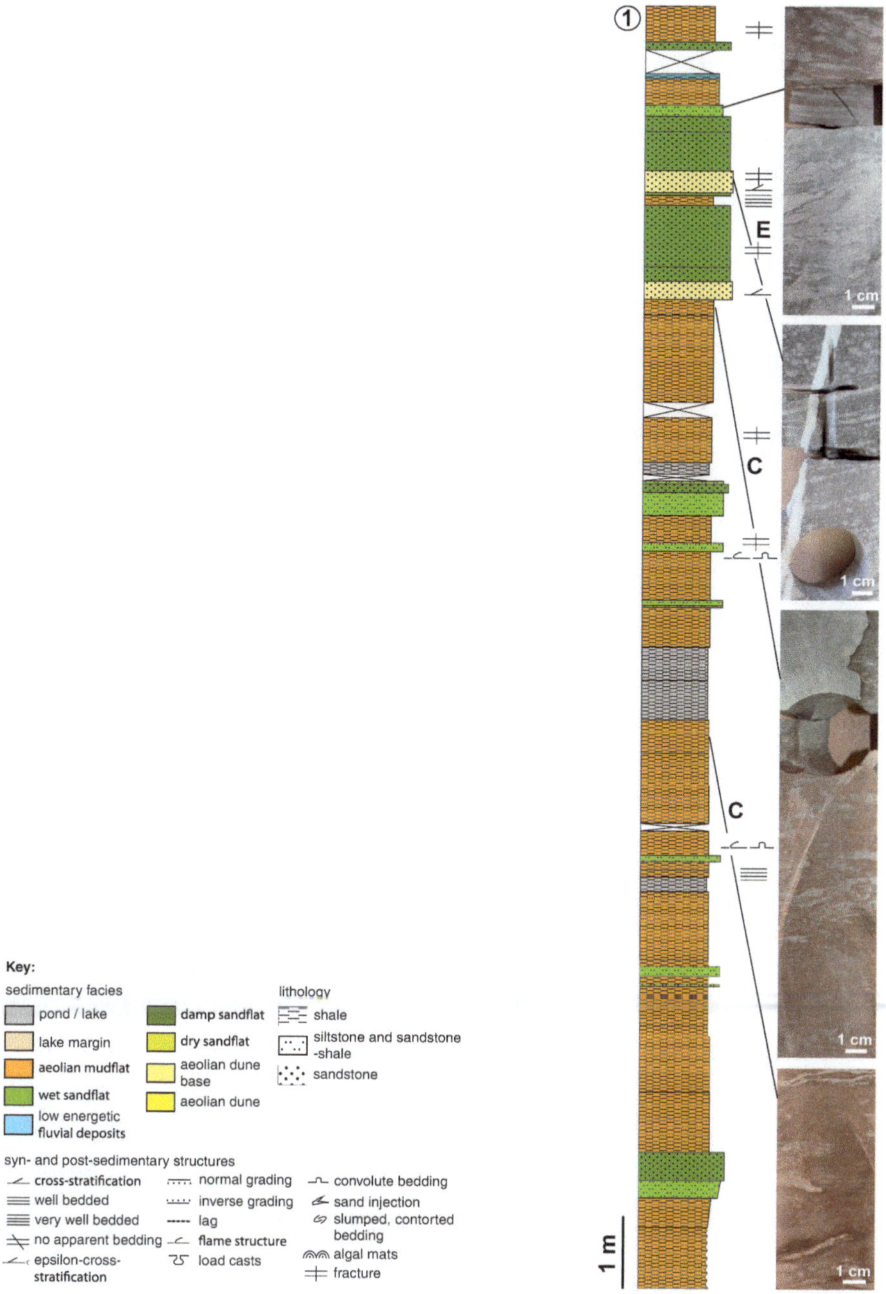

Fig. 5.18 Core *1*, Well 3. Photos of core successions plotted next to the lithology and depositional environment column. *C* and *E* mark photos of the features referred to in the text

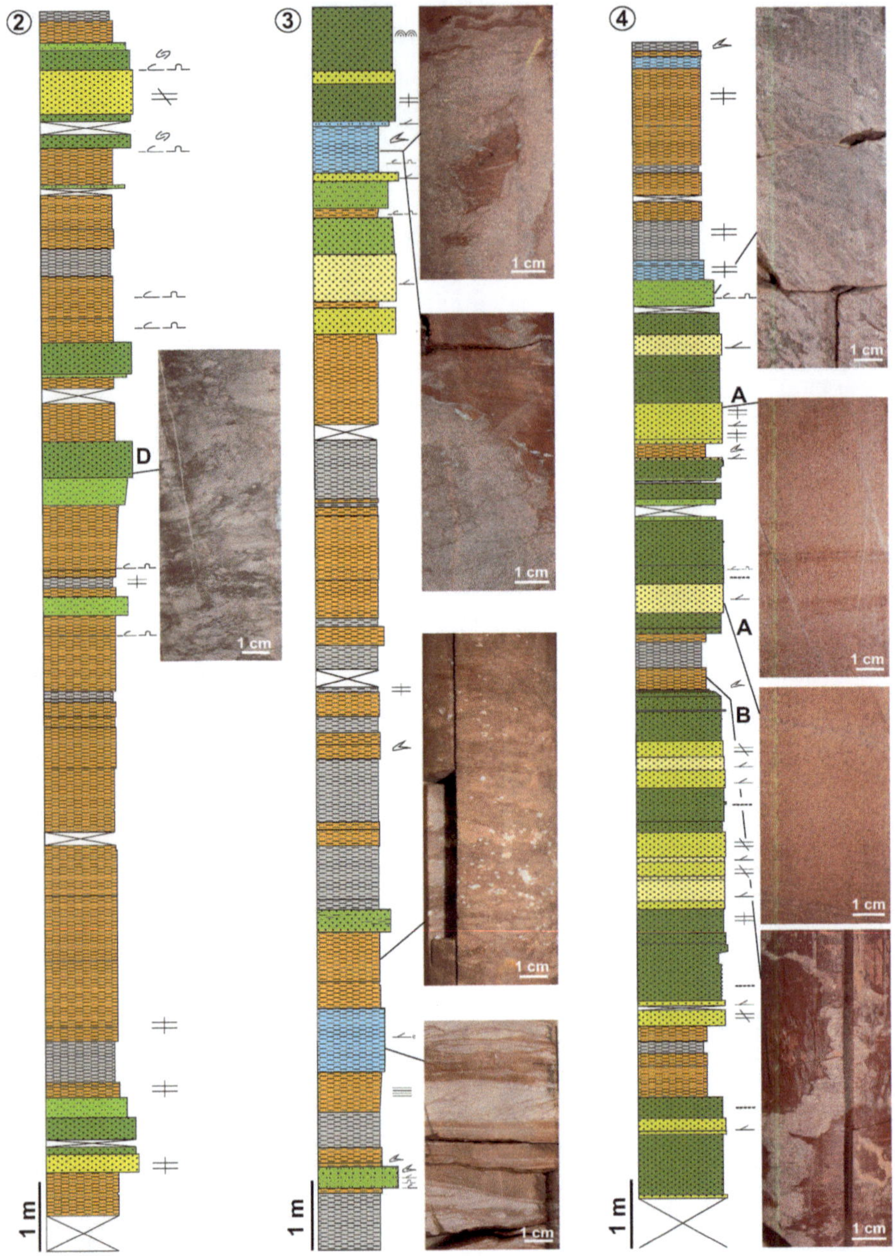

Fig. 5.19 Cores *2–4*, Well 3. Photos of core successions plotted next to the lithology and depositional environment column. *A–D* mark photos of the features referred to in the text

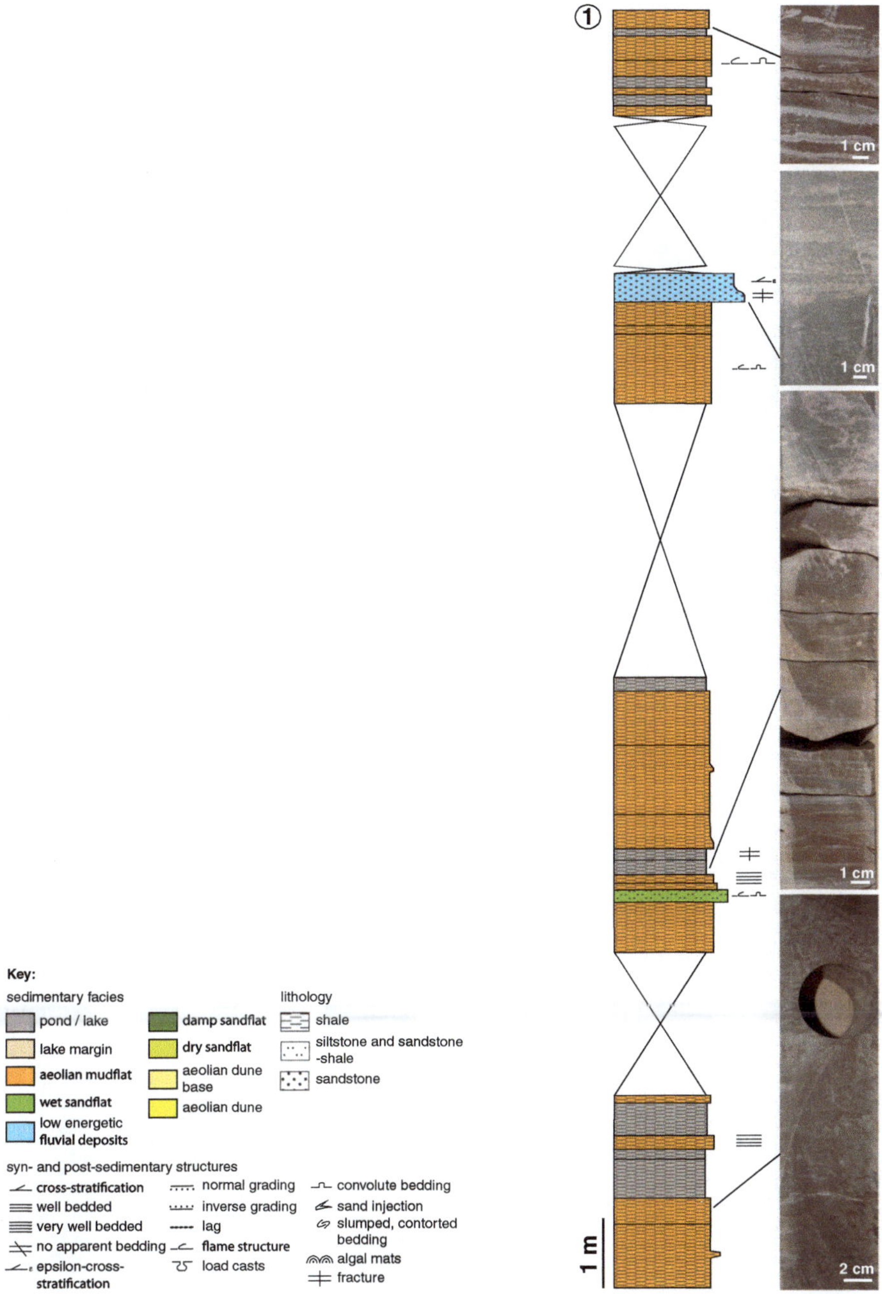

Fig. 5.20 Core *1*, Well 4. Photos of core successions plotted next to the lithology and depositional environment column

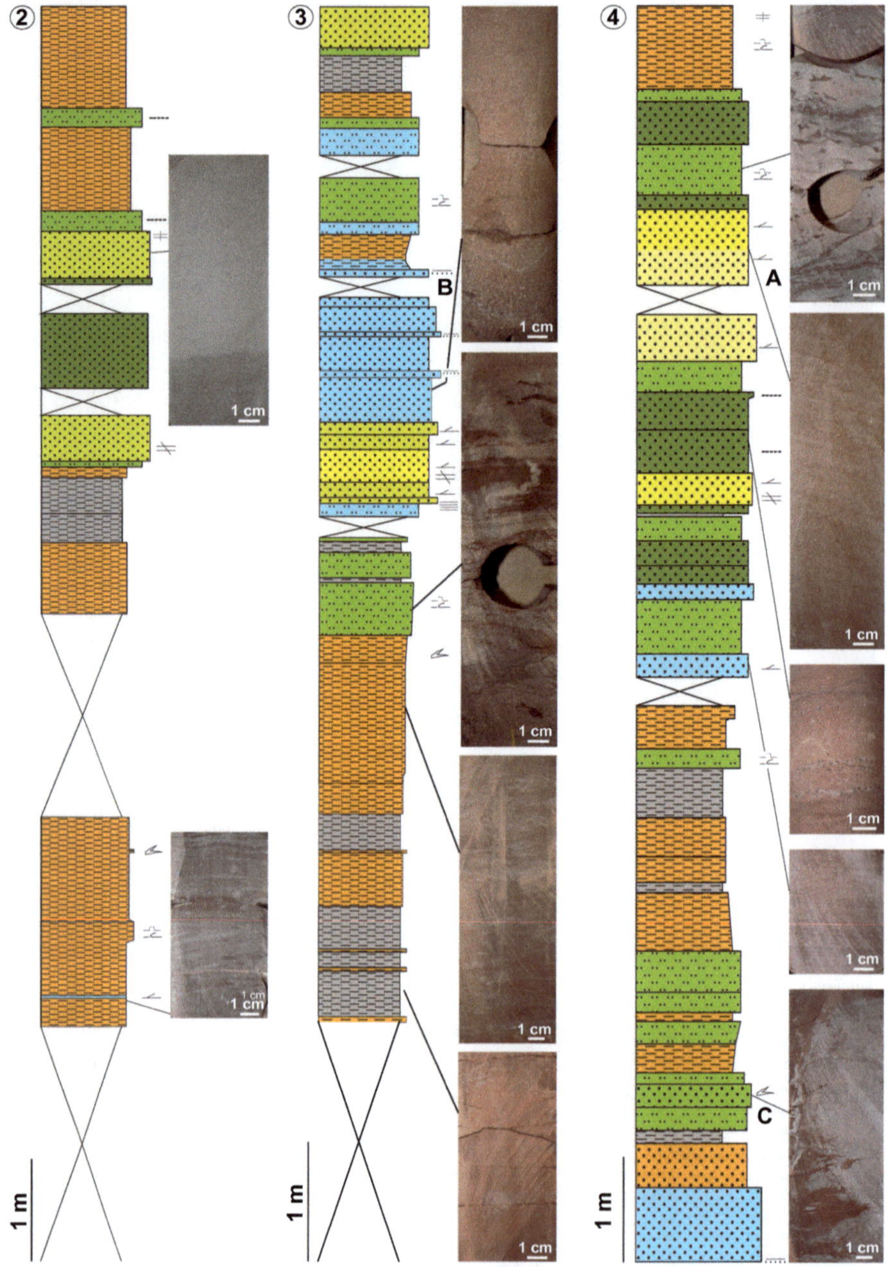

Fig. 5.21 Cores *2–4*, Well 4. Photos of core successions plotted next to the lithology and depositional environment column. *A* and *C* mark photos of the features referred to in the text

5.4 Interpretation

The interbedding of aeolian intervals with wet successions in the cores indicates that sedimentation mainly took place in wet environments. This is supported by the occurrence of several interdune deposits. Wet and damp sandflats further reveal that the groundwater table was shallow and located close to the sediment surface at all times. White mineralisation nodules, likely anhydrite and minor calcite, diagenetic concretions, and bleaching point to diagenetic overprinting of the aeolian intervals of the Munster Member (see Wells 2 and 3). Fractures are cemented by mica and quartz and in most cases offset sedimentary structures (cross bedding, layered deposits) of the Munster Member. Due to the limited thickness of dune successions, a differentiation of dune types throughout the wells is not possible. In general, the small thicknesses of aeolian successions in the cores indicate that dunes were also limited in height and did not exceed 30 m. All fluvial sediments identified in cores were deposited under low energy currents. Sheetflood deposits are the result of small temporary streams that discharged into lake or pond environments. Palaeotopographic gradients are assumed to have been very low due to the absence of erosional structures and a generally small grain size of the sediments. Sedimentation was mainly sub-aquatic as indicated by mudflats, pond and lake deposits. Damp and wet sandflat deposits developed at times of a shallow groundwater table or ephemeral flooding. Clay intervals often contain micro-fracture systems that are mainly layer-bound and horizontally oriented, while a vertical orientation is infrequently visible. Occasionally, salt efflorescence is observed along the fracture systems.

5.5 Comparison to the Study Area in East Frisia

The interest area discussed in this chapter is located ~200 km east of the East Frisian tight gas field (see Chaps. 3, 4, 6) and was located closer to the centre of the Southern Permian Basin during the Upper Rotliegend II. As a consequence, sedimentation is influenced by lake level changes of the great saline Rotliegend lake. In addition, the groundwater table is very shallow and located close to the sediment surface in the vicinity of the lake. The expansion of the great Rotliegend lake since the early Niendorf and marine ingressions during the Niendorf and Munster Members result in an increased clay and evaporite content in the deposited sediments. These facts are expressed in the predominant development of wet deposits (damp to wet sandflats, mudflats) that can be observed in the cores of the northern central German study area.

While core data from East Frisian focused on reservoir intervals of the Bahnsen and Wustrow, the here presented core analysis comprised the uppermost Members of the Hannover Formation (Bahnsen Member to Munster Member). In comparison, sedimentation in East Frisia was not directly influenced by the marine ingressions or changes of the Rotliegend lake level until the initial Zechstein flooding. There, deposition took place at the margin of the SPB. Due to limited sediment supply and high wind velocities, dunes were restricted in height. However, aeolian successions represent the dominant sedimentary facies in the cores [13].

The observations in the cores indicate that reservoir rock quality during the Bahnsen Member is better in East Frisia due to thicker aeolian successions and a lower clay content. However, it is likely that previous to the starting extension of the Rotliegend lake dry aeolian sediments dominated the deposition in northern central Germany and reservoir rocks of a higher quality are situated below the investigated interval. For the East Frisian area it can be assumed that the occurrence of aeolian successions decreases and wet deposits increase with a growing expansion of the Rotliegend lake or marine ingressions that took place at the latest with the onset of the Zechstein transgression. In general this study suggest that aeolian deposits and thus rocks of good reservoir quality can be found in younger intervals at the margin of the SPB compared to areas situated closer to the basin centre and the saline Rotliegend lake.

5.6 Conclusions

- The source area of the fluvial deposits is assumed to have been located in the south because the southernmost well (Well 4) shows the highest fluvial influence of all wells.
- All wells are located in an area, where aeolian sedimentation took place during the Bahnsen and partly during the Munster Member. Due to the palaeogeographic situation, just south of the great saline SPB lake, the sedimentation was influenced by a shallow groundwater table. Consequently, the dominant sedimentary facies in the aeolian successions are damp sandflats. Low energetic streams caused episodical, fluvial reworking. The influence of ephemeral lake high stands is expressed in the increased saline content in several core successions. The dominance of mudflats during the intermediate Members of the Hannover Formation points to a completely flooded and only ephemerally dry environment.
- In general, rocks of good reservoir quality are likely to be found in younger intervals at the margin of the SPB, compared to areas situated close to the basin centre and the saline Rotliegend lake.

References

1. Amthor JE, Okkerman J (1998) Influence of early diagenesis on reservoir quality of Rotliegende sandstone. Northern Netherlands. AAPG Bulletin 82(12):2246–2265
2. Bagnold RA (1954) The physics of blown sands and desert dunes, 2nd edn. Chapman and Hall, London
3. Gast R, Gaupp R (1991) The sedimentary record of the late permian saline lake in N.W. Germany. In: Renaut RW, Last WM (eds) Sedimentary and paleolimnological records of saline lakes. Natl Hydrol Res Inst, Saskatoon, Canada, 75–86
4. Gast R, Pasternak M, Piske J, Rasch H-J (1998) Das Rotliegend im nordostdeutschen Raum: Regionale Übersicht, Stratigraphie, Fazies und Diagenese. Geologisches Jahrbuch A 149:59–79
5. George GT, Berry JK (1993) A new palaeogeographic and depositional model for the upper rotliegend of the UK sector of the Southern North Sea. In: North CP, Prosser D J (eds), Characterization of fluvial and aeolian reservoirs. Geological Society, Special Publications, London, 73, pp 291–319
6. Glennie KW (1972) Permian rotliegendes of Northwest Europe interpreted in light of modern desert sedimentation studies. AAPG Bulletin 56(6):1048–1071
7. Glennie KW, Higham J, Stemmerik L (2003) Permian. In: The millennium atlas; petroleum geology of the central and northern North Sea. Geological society of London, London, 91–103
8. Legler B (2005) Faziesentwicklung im Südlichen Permbecken in Abhängigkeit von Tektonik, eustatischen Meeresspiegelschwankungen des Proto-Atlantiks und Klimavariabilität (Oberrotliegend, Nordwesteuropa). Schriftenreihe der Deutschen Gesellschaft für Geowissenschaften 47
9. Mountney NP, Jagger A (2004) Stratigraphic evolution of an erg margin aeolian system: The Permian Cedar Mesa Sandstone, SE Utah, USA. Sedimentology 51:713–743
10. Rieke H (2001) Sedimentologie, Faziesarchitektur und Faziesentwicklung des kontinentalen Rotliegenden im Nordostdeutschen Becken (NEDB). Ph.D. thesis, Potsdam University, Germany
11. Stollhofen H, Bachmann GH, Barnasch J, Bayer U, Beutler G, Franz M, Kästner M, Legler B, Mutterlose J, Radies D (2008) Upper Rotliegend to early cretaceous basin development. In: Littke R, Bayer U, Gajewski D, Nelskamp S (eds) Dynamics of complex intracontinental basins, The Central European Basin System. Springer, Berlin, pp 181–210
12. Strömbäck AC, Howell JA (2002) Predicting distribution of remobilized aeolian facies using sub-surface data: the Weissliegend of the UK Southern North Sea. Pet Geosci 8:237–249
13. Vackiner AA (2011) Sedimentary facies reconstruction and kinematic restoration of an upper permian tight gas field, north-western Germany. Ph.D. thesis, RWTH Aachen University, Germany
14. Ziegler PA (1982) Geological atlas of Western and Central Europe. Shell Internat Petrol Maatsch, The Hague

Chapter 6
Nano-Scale Porosity Analysis of a Permian Tight Gas Reservoir

6.1 Introduction

With growing energy needs and further exploitation of conventional hydrocarbon reservoirs, the importance of tight gas reservoirs has increased dramatically during the last years. However, porosity and permeability reducing processes in these reservoirs often remain enigmatic. Studies concerning seismic interpretation, seismic attribute analysis and well log information often reach a dead-end when closely spaced changes in the depositional and structural environments occur.

In this chapter a new approach of qualitative and quantitative reservoir petrology on samples from a tight gas field in the Upper Rotliegend II (Upper Permian) in Northern Germany is investigated. In a marginal position to the great Rotliegend playa lake in the Southern Permian Basin (SPB, Fig. 6.1), reservoir rocks of fluvio-aeolian origin were deposited under arid climate. During diagenesis, rocks have undergone an increased degradation of porosity and permeability triggered by migrating fluids [10, 9]. Enhanced cementation, such as an increased quartz overgrowth and secondary clay mineral crystallisation were the results of this porosity inverting processes, which were mainly recognized in aeolian (dune) deposits. The timing of this diagenetic effect and its relation to stress induced fluid pathways, e.g. fractures and faults are not fully understood.

Scanning electron microscopy-broad ion beam (SEM-BIB) analysis [2] and thin section analysis is applied for diagenetic observations, the quantification of pore symmetries and porosity estimation. Porosity and mineral content of a sample is analysed by image analysis software. The new technique overcomes the low resolution of standard optical microscopy when porosity and pore symmetries can not be quantified in areas where clay is the dominating mineral e.g. clay replacing feldspar mineral grains. To visualise and quantify porosity distribution and symmetry in such areas SEM-BIB images with the focus on inter-clay porosity are analysed. SEM-BIB results are upscaled, and porosity values and clay volumes derived from blue stained thin sections are calculated. Information about pore shapes and volumes can be used as input for fluid flow modelling. The aim of this study is to develop an automated workflow that provides reproducible results.

P. Antrett, *Characterization of an Upper Permian Tight Gas Reservoir*, Springer Theses, DOI: 10.1007/978-3-642-36294-1_6,

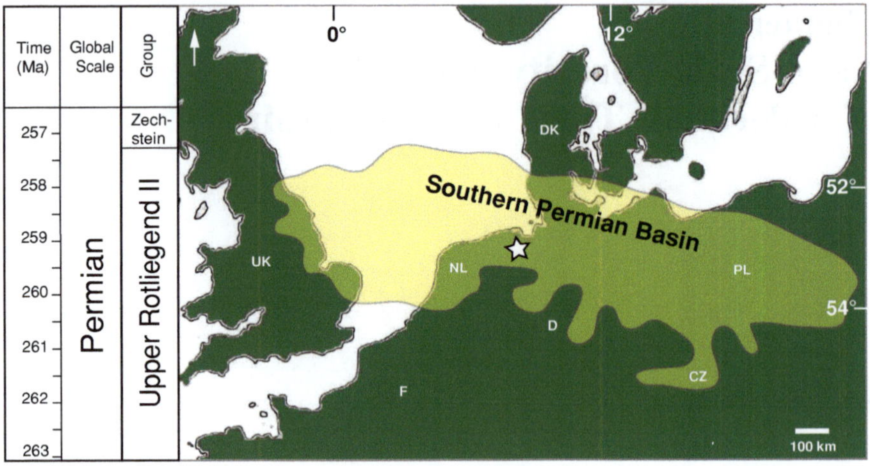

Fig. 6.1 Stratigraphic chart on the *left* (modified from [16], timescale according to Gast, 1995 [6]). Maximum extent of the Southern Permian Basin (modified after [22]). Study area marked by the *star*

6.2 Geologic and Sedimentary Setting

The W–E trending Southern Permian Basin (SPB) extends from the UK across the southern North Sea, covering northern Germany, Poland and the Baltic States, varying in width between 300 and 600 km. The basin centre was situated in Northern Germany. It reached its maximum extension during the Late Upper Rotliegend II, covering an area of $\sim$429,000 km^2. The tight gas field investigated is located in Upper Rotliegend II at the south-western part of the SPB in NW–Germany (Fig. 6.1). During the arid to semi-arid climate, sedimentation was dominated by siliciclastics and only minor evaporites [12]. Four major facies associations were identified by George and Berry [11], Strömbäck and Howell [21], Legler [16]: ephemeral fluvial (wadi), aeolian, sabkha and lacustrine environments. Strong aeolian winds supplied a huge amount of aeolian sediment from the east [8, 13, 18]. In addition, sediment was provided by ephemeral fluvial systems sourced by the Variscan Orogen and local highs in the south. Sedimentary facies analysis by Vackiner et al. [22] on cores of the investigated tight gas field reveals a close interbedding of dry aeolian sediments (dune, dune base and dry sandflats) with low clay contents with wet aeolian, interdune and lake/pond deposits (damp and wet sandflats and mudflats) with higher clay contents. The maximum dune interval in the cores is $\sim$2.5 m.

6.3 Data and Methods

The analysis is based on core material of 2 wells. In addition to optical microscopic analysis of blue stained thin sections, SEM-BIB samples with a very smooth surface were produced using a stand-alone argon ion beam cross section

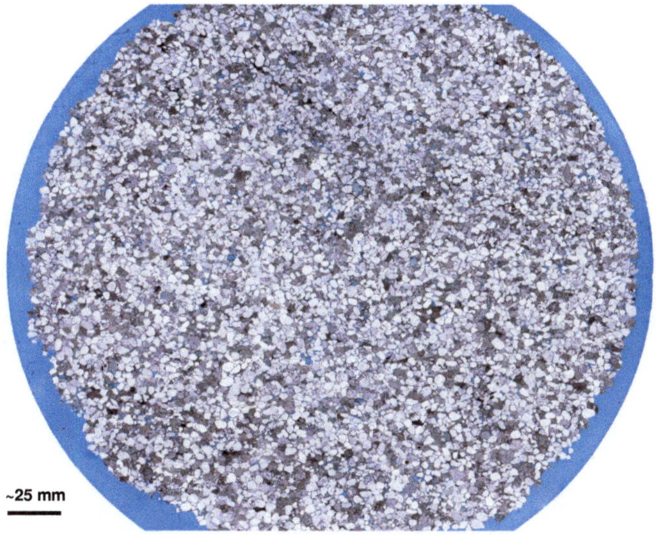

Fig. 6.2 Overview of a stitched thin section image. Figure composed of 180 individual images

polisher (JEOL SM-09010, Erdmann [4]) and a Zeiss Supra 55 electron microscope equipped with an EDX detector (EDAX, Apollo 10; [2]). Images were taken at a magnification of 3,000. SEM imaging studies were performed at 15 kV at a working distance of 10 mm to guarantee the consistent contrast that is required for the image stitching process and the threshold adjustment during image analysis. Thin sections were digitalised at 25 times magnification using the Leica DM 4,500 P polarisation microscope equipped with a Leica DFC 295 camera and a PC workstation with the IQ-Win Light software package (www.andor.com). Panoramic images of thin sections and SEM-BIB were compiled/stitched using the Autopano Giga 2.5 software by Kolor (Figs. 6.2 and 6.3; www.kolor.com).

Optical microscopic images consist of ~80–180 individual images, SEM-BIB panoramic images of ~340–400 individual images. The stitched images enable a continuous, dynamic zoom from the polished surface of 1.5 mm^2 down to pores of μm^2 size. For pore shape/symmetry and porosity analysis the open source software JMicroVision V.1.25 [19] was used. In blue stained thin sections, porosity and clay content were determined by adjusting a threshold manually. SEM-BIB images, backscatter electron microscope (BSE) images and EDX measurements were used for mineral identification. In a first step, clay minerals are identified and then isolated using standard image editing software. By manually adjusting a threshold in JMicroVision the inter-clay porosity was determined (Fig. 6.4). The total porosity of the sample was calculated by adding the porosity determined from the blue stained pores in the thin section to the inter-clay porosity from the SEM-BIB image (Fig. 6.5). The inter-clay porosity was calculated from the percentage of inter-clay porosity from the SEM-BIB image and the clay content of the thin section:

Fig. 6.3 Overview of a stitched SEM-BIB image. Figure composed of 361 individual images

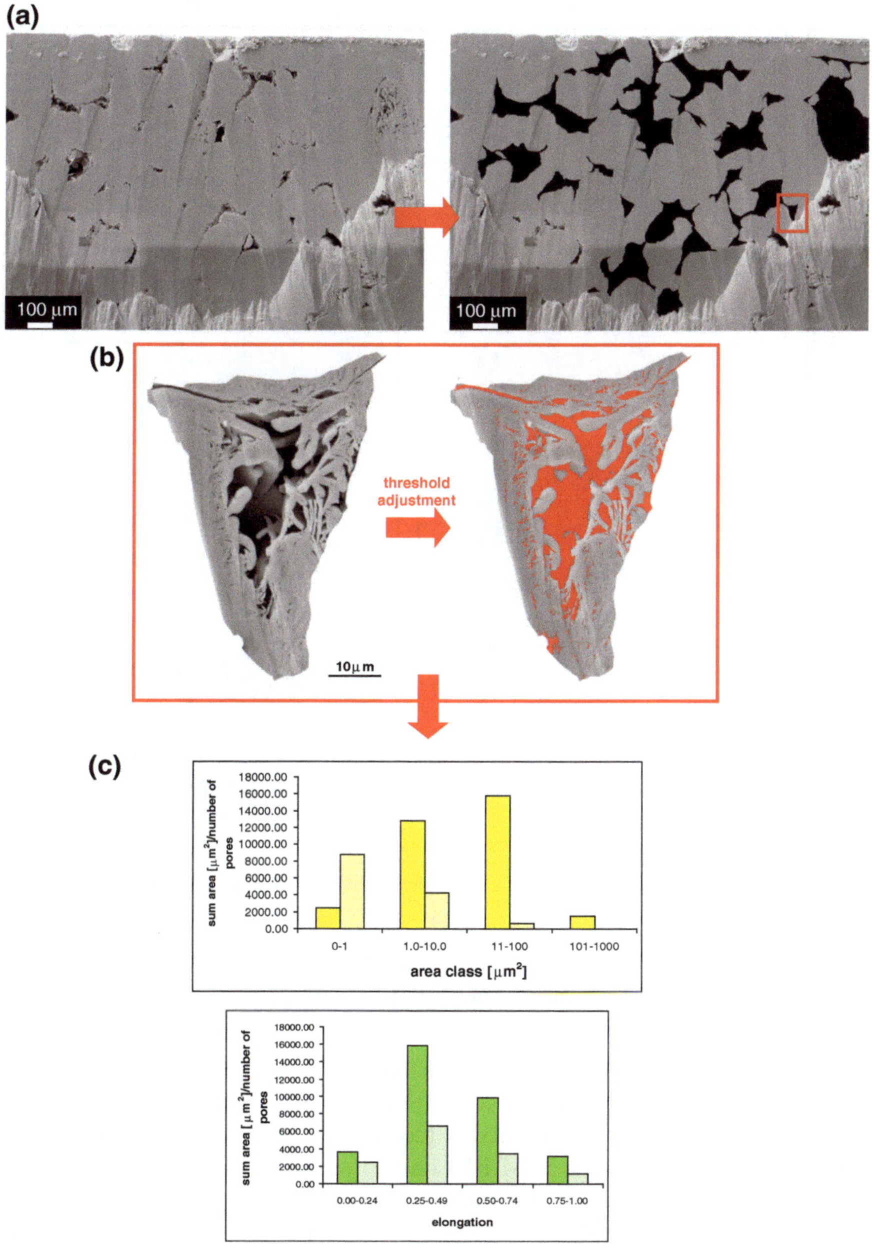

Fig. 6.4 Workflow for inter-clay porosity analysis. **a** identification and isolation of clay minerals. **b** Threshold adjustment to quantify inter-clay porosity. **c** Results of the pore analysis of the whole sample.

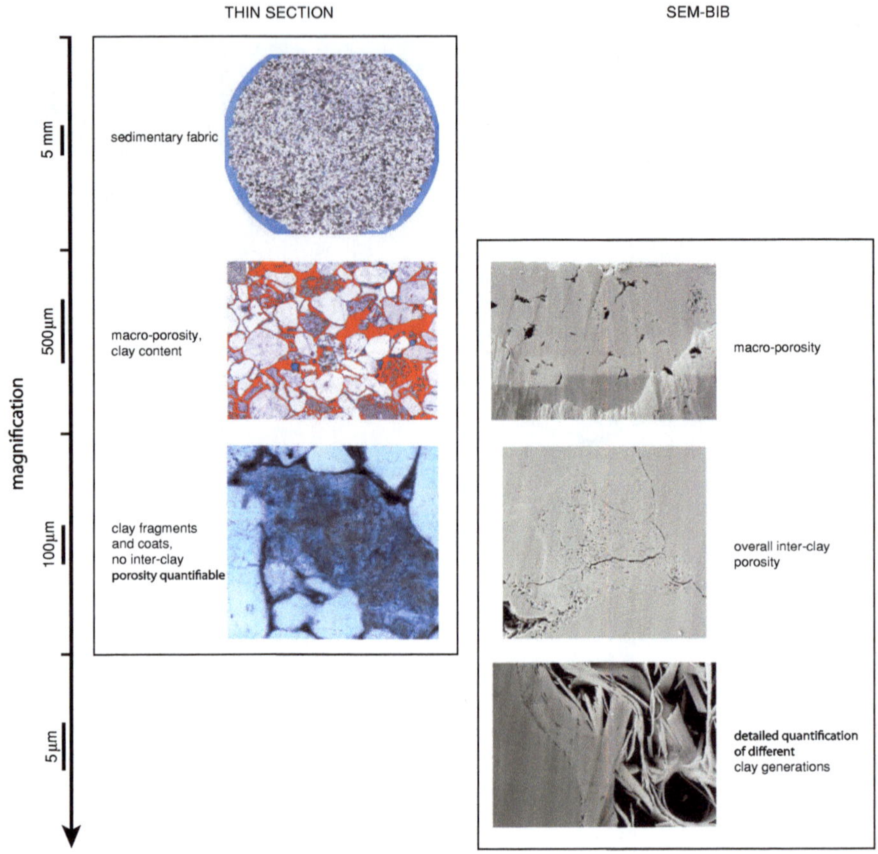

Fig. 6.5 Information available from thin sections and SEM-BIB images at different scales

$$\phi_{tot} = \phi_{ts} + \text{ ICP } \cdot \text{ CM}_{ts}$$

ϕ_{tot} = total porosity ϕ_{ts} = porosity in blue stained thin section ICP = inter-clay porosity. CM = clay mineral content in thin section

A minimum threshold of 10 pixel^2 (equals an area of 0.7 mm^2 in the thin section and 0.096 µm^2 in the SEM-BIB image) was established to eliminate pixel errors and to avoid false pore statistics.

Results of porosity analysis were compared to porosities derived from sonic velocities. In pore symmetry analysis, the pore area (2D) and the elongation (long axis vs. short axis) were determined. All the results directly provide the input information for fluid flow modelling.

6.4 Results

6.4.1 Diagenetic Observations, Pore Symmetry Quantification and Upscaling Approach

The SEM-BIB technique enables pore visualisation down to nm scale. The argon ion beam preparation enables unique insight into pore geometries and morphologies [2]. The number of individual pores per sample in the so far investigated 7 SEM-BIB images varies between ~20,000 and ~80,000. With the information provided by EDX and SEM-BIB analysis a relative timing of crystallisation was established (Fig. 6.6) and compared to existing analysis of diagenetic evolution of the Permian tight gas reservoir by Havenith et al. [14]. Especially with regard to different generations of clay mineral growth, these well polished samples provided valuable additional information.

Aeolian deposits show non-continuous illite and chlorite coatings around detrital quartz and feldspar grains (Fig. 6.7). Information about the timing of hydrocarbon migration was derived from euhedral titanite, growing on multiple generations of clay minerals (Fig. 6.7). In addition stress indicators were identified. Coarse chlorite meshwork in the pore space blocks/inhibits later illite growth and thus passively preserves initial porosity (Fig. 6.8).

In this study, an aeolian dune sample was analysed. Pore area quantification of the SEM-BIB image shows that the pore fraction with the smallest pore area with a pore size between 0.0961 μm^2 and 1 μm is most abundant (82 %, Fig. 6.4).

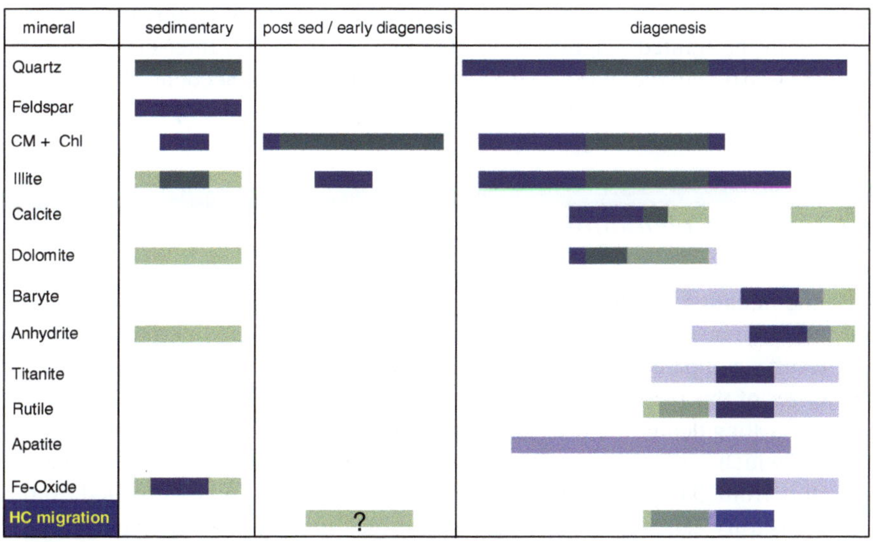

Fig. 6.6 Relative timing of crystallisation. Comparison of results of optical microscopy (*green*, [13]) and SEM-BIB (*blue*). Uncertainty displayed in *light colours*

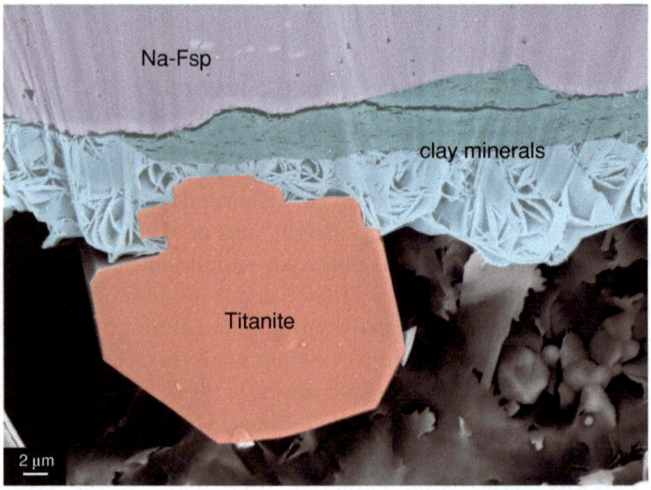

Fig. 6.7 Non-continuous clay coating of a detrital feldspar grain in *dark green*. The growth of euhedral *titanite* on two generations of *clay minerals* is associated with hydrocarbon migration

However, they only cover a relatively small area of 22 %. Pore symmetry was determined by the elongation and the ratio of long and short pore axis with a dominant elongation between 0.25 and 0.49 (Fig. 6.4).

In a second sample, an upscaling approach was applied combining information from thin sections and SEM-BIB. Colour threshold adjustment in the blue stained thin section gave a result of 0.9 % porosity and an amount of clay minerals including inter-clay porosity of 41.1 %. Inter-clay porosity is not further quantifiable with the limited magnification of optical microscopy. Inter-clay porosity in the SEM-BIB sample was 13.6 %, which taking into account the inter-clay porosity of 41.1 % in thin section, adds up to a total amount of porosity of 6.5 %.

6.4.2 Error Calculation

With the intention to quantify the error of manual threshold analysis on SEM-BIB images, three images were chosen in which all the pores were interpreted manually and results compared to the thresholding method. The first sample shows the challenges of this method applied on the 2D images. The SEM-BIB images provide a uniquely long depth of view into the pore that is partly filled by secondary quartz cement, which will also be identified by the threshold (Fig. 6.9). However, the upscaling from 2D sections in 3D only holds true if the pore area estimated from 2D sections is truly 2D. In general, the pore areas identified by automated interpretation are always underestimated [5]. Manually tracked pores gave a result of 11.5 ± 0.6 %, while threshold analysis is estimated 10.4 ± 0.1 % porosity for an

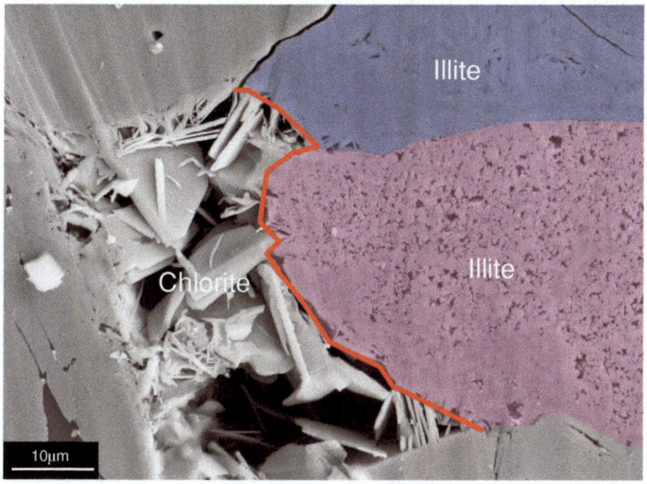

Fig. 6.8 Two illite generations. Further growth is blocked/inhibited by coarse chlorite meshwork

individual image (Fig. 6.9). Whereas the estimated porosity by calculation will be lower in an automated analysis, the induced fractures within SEM-BIB samples due to releasing stress from well recovery will increase the porosity by 0.6 % on the example of image 1 (Fig. 6.9). In two additional images manual interpretation versus threshold adjustment resulted in: 6.89 versus 6.08 % and 5.21 versus 4.47 %.

For the upscaling from SEM-BIB images to thin sections only the inter-clay pores are of interest, because pores of bigger sizes are quantifiable with thin section analysis. In addition, the 3D effect that is a challenge in the SEM-BIB images is absent in thin sections. Consequently, the major pore shown in image 1, which is not an inter-clay pore, is not included in SEM-BIB analysis. When the fractures are subtracted from the result of automated analysis the resulting porosity is 9.8 %. In comparison, the calculated porosity from sonic velocities is 9.4 ± 1.9 %.

6.5 Discussion and Conclusions

The information provided by the SEM-BIB method, in addition to optical microscopy, improves the diagenetic analysis. The nicely polished samples are a valuable source of additional information, especially concerning the differentiation of multiple clay mineral generations (Fig. 6.7). One example is the observation of non-continuous clay coats around detrital mineral grains, which are characteristic for active depositional settings [4, 1]. In such systems dunes are restricted in height as a result of high wind velocities and limited sediment supply [6]. Additionally, abraded clay coats indicate a further sediment transport [4, 1]. This observation

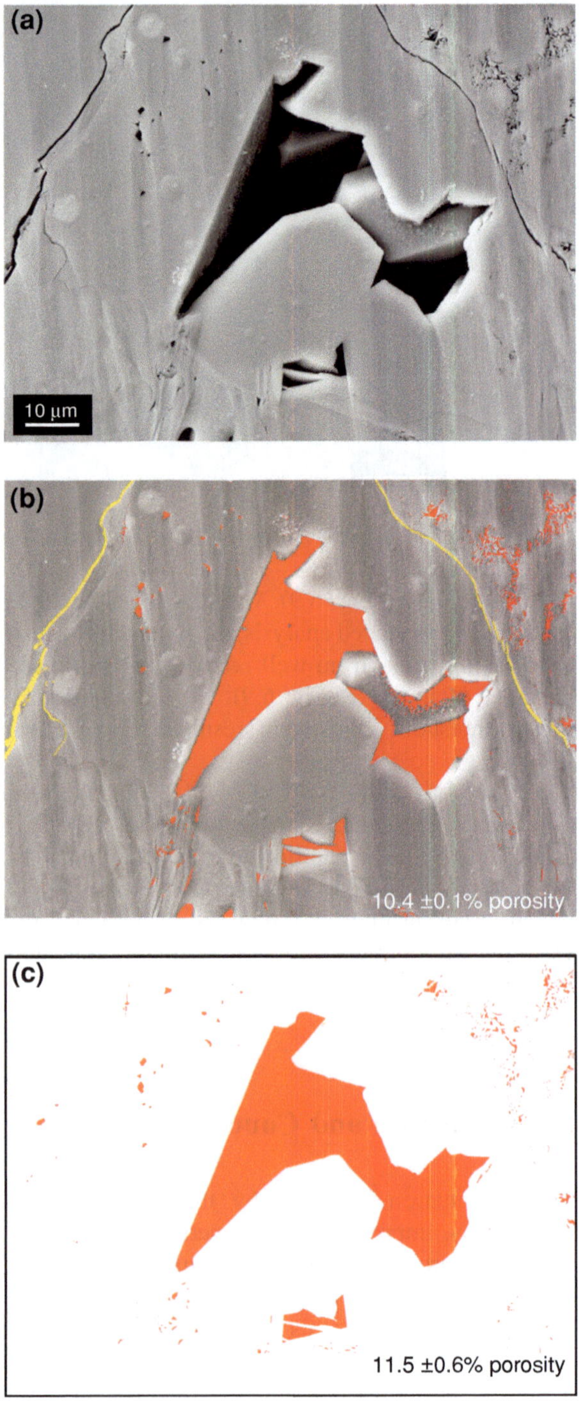

Fig. 6.9 **a** Original SEM-BIB image. **b** Result of manual threshold adjustment. Euhedral quartz in the pore is a challenge for threshold adjustment. Fractures highlighted in *yellow*. **c** Manual interpretation of porosity

supported previous core interpretations by Vackiner et al. [22] and is important concerning the identification of modern reservoir analogues.

Pores can be quantified with respect to their size, distribution, surface, roughness and mineralogical phases in tight reservoirs rocks. Those parameters serve as direct input for fluid flow modelling. The observations show that uncertainty decreases with increasing areas analysed. In this chapter, a minimum pore value/area was established for pore size interpretation to avoid false statistics and pixel errors. However, the charts (Fig. 6.4) indicate that the porosity in even smaller pores, which are not considered in this study, might only have a marginal influence on poro/perm characteristics. To investigate this assumption, fractal dimensions of pore size distributions as discussed by Krohn and Thompson [15] and Schlueter et al. [20] will be investigated in future studies.

The approach of upscaling from SEM-BIB to thin sections improves optical porosity estimation. Further upscaling to core and reservoir scale is investigated in the ongoing study. The application of the technique and the methodology introduced on the example of the tight gas field in Northern Germany can provide improved knowledge and understanding of the reservoir petrology in hydrocarbon reservoirs worldwide.

References

1. Adjukiewicz JM, Nicholson PH, Esch WL (2010) Prediction of deep reservoir quality using early diagenetic process models in the Jurassic Norphlet Formation Gulf of Mexico. AAPG Bulletin 94(8):1189–1227
2. Desbois G, Urai JL, Kukla PA, Konstanty J, Baerle C (2011) High-resolution 3D fabric and porosity model in a tight gas sandstone reservoir: a new approach to investigate microstructures from mm- to nm-scale combining argon beam cross-sectioning and SEM imaging. J Pet Sci Eng doi:10.1016/j.petrol.2011.06.004
3. Erdmann N, Campbell R, Asahina S (2006) Precise SEM cross-section polishing via Argon beam milling. Microscopy Today, 22–25 May 2006
4. Esch WL, Adjukiewicz JM, Reynolds AC (2008) Early grain-coat formation in chaco dune field, New Mexico: insight into formation mechanisms, distribution, and implications for predictive modeling to assist in deep play identification. Search and Discovery Article #50135, Posted October 29, 2008. Adapted from oral presentation at AAPG Annual Convention, San Antonio, TX, 20–23 April 2008
5. Fens TW (2000) Petro physical Properties from small rock samples using image analysis techniques. Published Dissertation. Technische Universiteit Delft
6. Fryberger SG, Dean G (1979) Dune forms and wind regime. In: McKee ED (ed.) A study of global sand seas. University Press of the Pacific, pp 137–169
7. Gast R (1995) Sequenzstratigraphie. In: Plein E (ed.) Stratigraphie von Deutschland I; Norddeutsches Rotliegendbecken. Rotliegend-Monographie Teil II. Courier Forschungs-Institut Senckenberg, Frankfurt a M, 183, pp 47–54
8. Gast R, Pasternak M, Piske J, Rasch H-J (1998) Das Rotliegend im nordostdeutschen Raum: Regionale Übersicht, Stratigraphie, Fazies und Diagenese. Geologisches Jahrbuch A 149:59–79
9. Gaupp R, Solms M (2005). Sedimentological and petrological investigations. In: Paleo oil- and gasfields in the Rotliegend of the North German basin: effects upon hydrocarbon

reservoir quality (Paläo-Öl- und Gasfelder im Rotliegenden des Norddeutschen Beckens: Wirkungen der KW-Migration auf die Speicherqualitäts-Entwicklung.) DGMK Research Report, 593–8: tight gas reservoirs—natural gas for the future. 1–1, 1–44
10. Gaupp R, Matter A, Platt J, Ramseyer K, Walzebuck JP (1993) Diagenesis and fluid evolution in deeply buried Permian (Rotliegende) gas reservoirs Northwest Germany. AAPG Bulletin 77(7):1111–1128
11. George GT, Berry JK (1993) A new palaeogeographic and depositional model for the upper rotliegend of the UK sector of the southern north sea. In: North CP, Prosser DJ (eds.) Characterization of fluvial and aeolian reservoirs, vol 73. Geological Society, London pp 291–319
12. Glennie KW (1972) Permian rotliegendes of northwest Europe interpreted in light of modern desert sedimentation studies. AAPG Bulletin 56(6):1048–1071
13. Glennie KW (1983) Early Permian (rotliegendes) palaeowinds of the North Sea. Sed Geol 34:245–265
14. Havenith VMJ, Meyer FM, Sindern S (2010) Diagenetic evolution of a tight gas field in NW Germany. DGMK/ÖGEW Frühjahrstagung 2010
15. Krohn CE, Thompson AH (1986) Fractal sandstone pores: automated measurements using scanning-electron-microscope images. Physical Review B 33(9):6366–6374
16. Legler B (2005) Faziesentwicklung im Südlichen Permbecken in Abhängigkeit von Tektonik, eustatischen Meeresspiegelschwankungen des Proto-Atlantiks und Klimavariabilität (Oberrotliegend, Nordwesteuropa). Schriftenreihe der Deutschen Gesellschaft für Geowissenschaften p 47
17. Plein E (1995) Stratigraphie von deutschland I: norddeutsches rotliegendbecken, rotliegend-monographie teil II, Courier Forschungs-Institut Senckenberg, Frankfurt. a. M, 183
18. Rieke H (2001) Sedimentologie, Faziesarchitektur und Faziesentwicklung des kontinentalen Rotliegenden im Nordostdeutschen Becken (NEDB). Ph.D. thesis, Potsdam University, Germany
19. Roduit N (2011). JMicroVision: image analysis toolbox for measuring and quantifying components of high-definition images. Version 1.2.7. http://www.jmicrovision.com. Accessed May 2011
20. Schlueter EM, Zimmermann RW, Witherspoon PA, Cook NGW (1997) The fractal dimension of pores in sedimentary rocks and its influence on permeability. Eng Geol 8:199–215
21. Strömbäck AC, Howell JA (2002) Predicting distribution of remobilized aeolian facies using sub-surface data: the Weissliegend of the UK Southern North Sea. Pet Geosci 8:237–249
22. Vackiner A, Antrett P, Back S, Kukla P (2010). Sedimentological and structural analysis of a Rotliegend tight gas field in NW Germany. GMK/ÖGEW-Frühjahrstagung 2010, Fachbereich Aufsuchung und Gewinnung, Celle, 12–13
23. Ziegler PA (1982) Geological atlas of western and central Europe. Shell Internat Petrol Maatsch, The Hague

Chapter 7
Conclusions and Outlook

7.1 Résumé

Dealing with the question to which extent seismic data can provide crucial information about the architecture and the framework of tight gas reservoirs, an intensive analysis of a 3D seismic data set of the Upper Rotliegend (Upper Permian) in East Frisia in Northern Germany was carried out. The main focus lay on seismic attribute analysis, which revealed a polygonal pattern of different scales on time-slices respectively horizon slices. In vertical seismic sections, the majority of the outlines of the polygonal pattern were identified as faults. Offsets of these faults varied from hundreds of metres to some tens of metres. However, in some cases no offset was visible at all, most likely as a result of limitations in vertical resolution of seismic data. In order to gain a better understanding of the recognized features, a suitable modern analogue was identified with the Panamint Valley in California, United States. Both study areas have a similar structural and sedimentary setting. Sediments were deposited under arid climate in an extensive structural setting with a strike slip component. The dominant sedimentary facies recognized during the field study in the Panamint Valley and from core material from Germany are dry aeolian (dunes and dry sandflats) and wet aeolian (damp and wet sandflats and mudflats) deposits.

On satellite images of the Panamint Valley, mega-crack systems were indentified on the ephemeral dry lake in the centre of the northern part of the Panamint Valley. The dimensions of these features reach lengths of more than a kilometre and depths of more than one metre. They form a polygonal pattern consisting of multiple lineaments of different dimensions, from km-scale to m-scale. Ground resistivity measurements of these mega-cracks implicated a vertical influence of the crack-, respectively fissure system of more than 3.5 m and in a case described by Langford and Eshete [7], even up to more than 8 m. Although a tectonic component has to be taken into consideration for the development of the mega-cracks in an active region like the Basin and Range Province, their evolution is mainly triggered by multiple wetting and drying cycles of (predominantly) clay minerals.

P. Antrett, *Characterization of an Upper Permian Tight Gas Reservoir*, Springer Theses, DOI: 10.1007/978-3-642-36294-1_7,

It can be summarised in four major steps: (1) The development of subsurface cavities, which grow towards the playa until the surface finally collapses. (2) The cracks are then refilled predominantly by wind-blown sand and collapsing crack walls. Vegetation in the cracks serves as sediment trap. (3) Sandy bulges develop around the bushes and during time get eroded after vegetation died. In contrast to the surrounding non-cracked playa surface, the areas of refilled cracks are composed of coarser grained sediment. (4) Due to the established heterogeneity in the deposited sediments, future cracks preferably develop at locations of refilled cracks. As a result the vertical influence of these features increases continuously.

Increased stress during burial and differences in compaction between heterogeneous and homogeneous areas on the former playa surface favour the development of weakness zones [2]. Due to better porosity, as a result of a higher amount of coarser grained sediment as crack-fills, these zones represent pathways for fluid migration. With time, some of these zones are likely to develop into faults or even into a polygonal fault system as described by Cartwright et al. [2].

The geological model, developed on the modern analogue, combined with the hypothesis of Cartwright et al. [2] was then applied to the interest area of the Permian tight gas field in Northern Germany. Results revealed that the polygonal patterns of the two study areas were geometrically and sedimentologically comparable. Based on these observations, it is conceivable that the fault and fracture pattern, identified by seismic attribute analysis, initially developed from a sedimentary setting similar to the Panamint Valley. Concerning reservoir quality, such fault systems including the subtle sub-seismic faults can have an influence on fluid migration and thus on diagenetic processes. In some cases the outlines on the z-slices are associated with depressions in the vertical sections affecting 3–4 seismic reflectors. Comparable features were interpreted as pipes (fluid migration pathways) by Hustoft et al. [6] from the Miocene Kai-Formation of the mid-Norwegian margin. In the case of the studied reservoir in Northern Germany, a decrease of initially good reservoir quality, especially in aeolian dune deposits, is described by Gaupp et al. [5] and Gaupp and Solms [4] as a result of fluid migration. Porosity inversion decreases with increasing distance to neighbouring faults. The fracture and fault system is likely to be responsible for subtle reservoir compartmentalisation in the study area in East Frisia, Germany.

In addition, a sedimentary facies analysis of the cores from the East Frisian tight gas field, with a focus on sedimentary facies interpretation, was carried out. The workflow for the differentiation of sedimentary facies was mainly based on grain sizes and internal structures/geometries. To compare the sedimentary facies distribution and depositional environment of the tight gas field with a second interest area in northern central Germany, ~200 km east of the tight gas field, 4 additional cores were analysed. In the second study area, the majority of aeolian sediments occur in the Wustrow, Bahnsen and Munster Member and were deposited in a temporarily wet environment. The limited thickness of the dune intervals did not allow a distinct differentiation in various dune types, as it was possible in the East Frisia area. Fluvial sediments developed under low energy currents. Sedimentation mainly took place in a sub-aquatic milieu. Clay intervals in the cores often contain micro-fracture systems that are, in most of the cases, layer bound with a minor vertical component. This information might be essential for future field development, especially with the focus on hydraulic fracturing.

In contrast to the sedimentary facies analysis of the cores from East Frisia, core material of the Bahnsen and Munster Members, which directly overly the actual reservoir interval, was analysed in the northern central German study area. Palaeogeographically, the northern central German gas field is located closer to the centre of the SPB than the East Frisian interest area. Core material suggests a direct influence of the lake level changes of the great saline lake of the SPB and marine ingressions on the sedimentation of the northern central German study area. As a result, reservoir rock quality during the Bahnsen Member is better in East Frisia due to the occurrence of thicker aeolian successions and lower clay contents. For the East Frisian area, it can be assumed that aeolian successions decrease and wet deposits increase with a growing expansion of the Rotliegend lake or marine ingressions that took place at the latest with the onset of the Zechstein transgression. In general, this study suggest that aeolian deposits and thus rocks of good reservoir quality can be found in younger intervals at the SPB margin compared to areas situated near the basin centre and the saline Rotliegend lake.

7.2 General Implications

This study indicates that seismic data provides essential information about reservoir properties and improve reservoir modelling and field development (Chap. 3). However, the establishment of a detailed geological model requires additional information besides seismic data. The high importance of modern analogues for the development of the structural and depositional model, especially with the focus on tectonic and sedimentary interaction can not be underestimated. Only with the development of these models, an explanation of the origin of the features recognized on seismic data is possible. It enables a distinct characterization of the interest area, and thus allows an integration of the information which results in a refinement of the geological models (Chaps. 3 and 4). For future exploration, extensive seismic analysis, including multiple amplitude-dependent and amplitude-independent seismic attributes, is recommended. The focus of investigation should lie on z-slices (respectively time slices) in addition to the vertical sections. Furthermore, additional information about diagenesis, porosity values, porosity distribution/classification can be provided by SEM-BIB analysis, which supports and extends thin sections analysis and builds the foundation for an upscaling workflow (Chap. 6) that improves the understanding of reservoir properties.

Results of this thesis emphasise the importance of an improved understanding of the complex reservoir architecture of hydrocarbon reservoirs. This includes the heterogeneity in the sedimentary succession in addition to subsurface data, well logs and core data. Although time consuming, only intensive multidisciplinary, multiscale studies and the analysis of modern analogues, can provide suitable input for risk assessment and minimize exploration risks.

7.3 Outlook

Chapter 6 provided an approach for the upscaling of SEM-BIB images to thin sections as a first step in porosity extrapolation. Future work will focus on establishing a workflow for further upscaling of porosity from thin section to plug, from plug to core and from core to reservoir dimensions. Additionally, SEM-BIB images and thin section analysis revealed non-continuous clay coatings around detrital quartz and feldspar grains. In general, grain coats are interpreted to develop very early during diagenesis, or even during the sedimentation process in the lower, humid parts of dunes. The abrasion recognized, indicates a further sediment transport in an active dune system, in which dune heights are limited due to high wind velocities and limited sediment supply. Continuous clay coats are thought to indicate a stable, stabilized dune system allowing the development of higher dunes [3, 1]. However, processes of clay coat development are not entirely understood. Yet, clay coats have a high influence on later diagenetic processes during burial. Continuous coats around quartz grains prevent quartz overgrowth during diagenesis and as a result prevent porosity reduction. For a better understanding of these earliest diagenetic processes, samples from modern analogues of different dune types (barchanoid, aklé and transverse dunes) and heights are required. Additionally, the dune samples should be analysed concerning their position in the dune field and in the individual dune, e.g. windward side or lee side, dune base or dune ridge, coarse grained sheet flow or fine grained suspension fallout. Proposed study locations are the Death Valley National Park in California, United States, the Namib Desert in Namibia and the Wahiba Sands in Oman. In the Death Valley National Park, a variety of dune types can be analysed, ranging from small barchanoid dunes of 1–5 m height, to amalgamated star dunes of more than 200 m in Eureka Valley. Clay coats, covering detrital grains, were noticed in samples from the Death Valley National Park during this study. In the Namib desert and the Wahiba sands, grain coats can be analysed, especially with the focus on lateral distribution dependant on the distance to wet or moist environments as source area of the clay minerals. In addition, the influence of an increased carbonate fraction on grain coat development can be observed in the Wahiba Sands in the Oman desert. In Namibia, in addition to different dune types (barchanoid dunes at the Skeleton Coast and huge transverse dunes in the inland Namib), differences between red Fe-oxide coats and clay coats are of interest. Finally, the information gained from this analysis will increase accuracy of the upscaling process and provide input for diagenesis simulations.

References

1. Adjukiewicz JM, Nicholson PH, Esch WL (2010) Prediction of deep reservoir quality using early diagenetic process models in the Jurassic Norphlet formation Gulf of Mexico. AAPG Bull 94(8):1189–1227
2. Cartwright J, James D, Bolton A (2003) The genesis of polygonal fault systems: a review. In: Van Rensbergen P, Hillis RR, Maltman AJ, Morley CK (eds) Subsurface sediment mobilization. Geological society, vol 216. Special Publications, London, pp 223–243
3. Esch WL, Adjukiewicz JM, Reynolds AC (2008) Early grain-coat formation in chaco dune field, New Mexico: insight into formation mechanisms, distribution, and implications for predictive modeling to assist in deep play identification. Search and discovery article #50135, posted 29 October 2008. Adapted from oral presentation at AAPG Annual Convention, San Antonio, 20–23 April 2008
4. Gaupp R, Solms M (2005) Sedimentological and petrological investigations. In: Paleo Oil- and Gasfields in the Rotliegend of the North German Basin: Effects upon Hydrocarbon Reservoir Quality (Paläo-Öl- und Gasfelder im Rotliegenden des Norddeutschen Beckens: Wirkungen der KW-Migration auf die Speicherqualitäts-Entwicklung.) DGMK Research Report, 593–598: Tight Gas Reservoirs—Natural Gas for the Future. 1–1, 1–44
5. Gaupp R, Matter A, Platt J, Ramseyer K, Walzebuck JP (1993) Diagenesis and fluid evolution in deeply buried Permian (Rotliegende) gas reservoirs, northwest Germany. AAPG Bull 77(7):1111–1128
6. Hustoft S, Mintert J, Bünz S, Nouzé H (2007) High-resolution 3D-seismic data indicate focussed fluid migration pathways above polygonal fault systems of the mid-Norwegian margin. Mar Geol 245:89–106
7. Langford RP, Eshete T (2000) Fissure behaviour in the Chihuahuan desert, and depth estimation using sediment strength. Environ Eng Geosci 4(4):301–309

MIX
Papier aus verantwortungsvollen Quellen
Paper from responsible sources
FSC® C105338

If you have any concerns about our products,
you can contact us on
ProductSafety@springernature.com

In case Publisher is established outside the EU,
the EU authorized representative is:
Springer Nature Customer Service Center GmbH
Europaplatz 3, 69115 Heidelberg, Germany

Printed by Libri Plureos GmbH
in Hamburg, Germany